I0008618

Considering Use and Design of Speech Recognition Systems

Ben Kraal

Considering Use and Design of Speech Recognition Systems

Investigating Users of Complex Socio-Technical Systems

VDM Verlag Dr. Müller

Imprint

Bibliographic information by the German National Library: The German National Library lists this publication at the German National Bibliography; detailed bibliographic information is available on the Internet at http://dnb.d-nb.de.

   Any brand names and product names mentioned in this book are subject to trademark, brand or patent protection and are trademarks or registered trademarks of their respective holders. The use of brand names, product names, common names, trade names, product descriptions etc. even without a particular marking in this works is in no way to be construed to mean that such names may be regarded as unrestricted in respect of trademark and brand protection legislation and could thus be used by anyone.

Cover image: www.purestockx.com

Publisher:
VDM Verlag Dr. Müller Aktiengesellschaft & Co. KG , Dudweiler Landstr. 125 a, 66123 Saarbrücken, Germany,
Phone +49 681 9100-698, Fax +49 681 9100-988,
Email: info@vdm-verlag.de

Zugl.: Canberra, University of Canberra, 2006

Copyright © 2008 VDM Verlag Dr. Müller Aktiengesellschaft & Co. KG and licensors
All rights reserved. Saarbrücken 2008

Produced in USA and UK by:
Lightning Source Inc., La Vergne, Tennessee, USA
Lightning Source UK Ltd., Milton Keynes, UK
BookSurge LLC, 5341 Dorchester Road, Suite 16, North Charleston, SC 29418, USA

ISBN: 978-3-639-06758-3

Acknowledgements

Academically, there are many people to thank. My supervisors, Michael Wagner, Penny Collings and Anni Dugdale are, above all, patient. It was Michael who led me to automatic speech recognition as an application area to study from the point of view of interaction design. He introduced me to Penny Collings who brought her wealth of experience in HCI to my supervision team. When it became clear that I was going down the path of using social methods to conduct much of my research it was Penny who introduced me to Anni Dugdale, sociologist and Actor-Network theorist. All of my supervisors were free with their time and rich with advice and guidance. Since the first day I arrived at the University of Canberra they have treated me with much respect.

I must thank the anonymous interview subjects who gave up their work time and allowed me to ask questions about their work, their lives and their injuries. I'd especially like to thank Sue Woodward for making it possible for me to meet my interview subjects. I'd like to thank the Hansard Editors who gave up their work time to tell me about their work and how it was done.

All the people at the ACT Magistrates Court were very generous with their precious time. Ron Cahill, Chief Magistrate (among other duties) was remarkably accessible for someone so busy and his staff and their colleagues were unfailingly happy for me to ask, what must have been to them, simple questions about their work and how they went about doing it.

My parents and sister were very supportive when I left the city of my birth to travel interstate to study. My inability to explain my research clearly left them in the dark as to what I was doing for the best part of four years.

Finally, I'd like to thank my wife, Marice for all her love and support, both emotional and financial, during the time I've taken to produce this thesis. She shared in my small victories and commiserated my defeats. Young Mr. Daniel, who came along towards the end, provided much distraction and laughter. I love them both.

List of Publications

1. Ben Kraal, Michael Wagner, and Penny Collings. Improving the design of dictation software. In *Australian Speech Science and Technology Conference*, University of Melbourne, Victoria, Australia, November 2002. ASSTA.

2. Ben Kraal, Penny Collings, Anni Dugdale, and Michael Wagner. An ethnography of speech recognition. In *OZCHI 2004*, University of Wollongong, NSW, November 2004. CHISIG.

Abstract

Talking to a computer is hard. Large vocabulary automatic speech recognition (ASR) systems are difficult to use and yet they are used by many people in their daily work. This thesis addresses the question: How is ASR used and made usable and useful in the workplace now?

To answer these questions I went into two workplaces where ASR is currently used and one where ASR could be used in the future. This field work was done with designing in mind. ASR dictation systems are currently used in the Australian Public Service (APS) by people who suffer chronic workplace overuse injuries and in the Hansard department of Parliament House (Hansard) by un-injured people.

Analysing the experiences of the users in the APS and at Hansard showed that using an ASR system in the workplace follows a broad trajectory that ends in the continued effort to maintain its usefulness. The usefulness of the ASR systems is 'performed into existence' by the users with varying degrees of success. For both the APS and Hansard users, they use ASR to allow work to be performed; ASR acts to bridge the gap between otherwise incompatible ways of working.

This thesis also asks: How could ASR be used and made usable and useful in workplaces in the future? To answer this question, I observed the work of communicating sentences at the ACT Magistrates Court.

Communicating sentences is a process that is distributed in space and time throughout the Court and embodied in a set of documents that have a co-ordinating role. A design for an ASR system that supports the process of communicating sentences while respecting existing work process is described.

Moving from field work to design is problematic. This thesis performs the process of moving from field work to design, as described above, and reflects the use of various analytic methods used to distill insights from field work data.

The contributions of this thesis are:

- The pragmatic use of existing social research methods and their antecedents as a corpus of analyses to inspire new designs;

- a demonstration of the use of Actor-Network Theory in design both as critique and as part of a design process;

- empirical field-work evidence of how large vocabulary ASR is used in the workplace;

- a design showing how ASR could be introduced to the rich, complicated, environment of the ACT Magistrates Court; and,

- a performance of the process of moving from field work to design.

Contents

List of Figures

CHAPTER 1

Introduction

This thesis is about designing automatic speech recognition[1] systems for people to use; this thesis is also about usability and how usability comes about. Usability is often conceived of as something that is a property of an object or, more often, piece of software. Even more often it falls to designers to instil "usability" in an object by discovering what it is that "users" want and how they react to various elements of a piece of software, usually through conducting usability tests. This thesis is about how usability is not a property of an object but is co-produced through the interaction of many elements both human and non-human, social and technological. Instead of seeing usability as a property that is designed into a software product, this thesis takes the view that usability is the result of a complex interplay of human and non-human elements that include:

- front-line users;

- the particular software product in use;

- the interaction of other software products with the product in use; and,

- other human and non-human actors in the user's vicinity.

---

[1]Briefly, speech recognition is what people do. *Automatic* speech recognition (ASR) is what computers do.

1

This thesis does not seek to diminish the role that designers have to play in the creation of usable products. The progress that has been made in raising the profile of usability has made the gap between usable and less-usable technologies even more apparent.

Automatic speech recognition (ASR) is a field of research that is multi-disciplinary. Linguists, engineers, computer-scientists, designers and psychologists work together to build computer systems that can recognise human speech. The ASR research effort is usually concentrated in the laboratory but automatic speech recognition software products are increasingly found in the real world and are encountered daily by many people. The usability of ASR systems is thought to depend largely on their recognition accuracy (Huang et al., 1999) because a system with a low recognition accuracy, that is one that incorrectly recognises many words, is much more difficult to use than one that has a high recognition accuracy.

There are many kinds of ASR application. ASR is used in telephony applications (Yankelovich, 1996; Shriver and Rosenfeld, 2002) to allow people to speak instead of press buttons, replacing touch-tone menus. It is used in factories (Rollins et al., 1983) where it allows workers to have the use of both their hands and interact with a computer at the same time. It is used in cars (Lockwood and Boudy, 1992; Heisterkamp, 2001) so that drivers may keep their eyes on the road and their hands on the wheel while interacting with navigation systems. And it is used in large vocabulary dictation systems often on desktop computers(for example Karat et al., 2000). This thesis is concerned with large vocabulary systems.

There are two stories contained in this thesis. The first story is that of users of large vocabulary dictation systems. The second is that of the design of an ASR system for the Australian Capital Territory Magistrates Court. These stories are tied together in the next section.

1.1   Research Origins

This research began when the ACT Magistrates Court (the Court) discussed with the University of Canberra the possibility of implementing an ASR system for the Magistrates to use during the process of sentencing.

In the Court, the magistrates use rubber stamps to speed up the process of writing judgments, or sentences, on the bench. Many judgments are not encompassed by the stamps which requires writing out a full judgment in long-hand which is a time consuming process. Because these "judgments" are often not final but are often steps towards a final judgment that closes the case, I have termed them "outcomes".

Determining outcomes, as I have come to understand it, is a highly charged moment in the Court when the magistrate speaks an outcome for the case that he or she is hearing. An outcome may be a sentence, for example a fine or jail term or it may be the decision to set a case over to allow all the parties to the case more time to gather relevant information. An outcome may also be a procedural decision specific to the Court such as a request by the magistrate for any number of specialised reports that are used to inform the actual sentence when it is finally delivered.

The question I asked when I began this thesis was: What form could ASR at the Court take?

Initially, I had taken a software development model as the model of this thesis. Having acquired a client, the Court, I determined to follow user-centered design principles (see, for example, Shneiderman, 1992; Nielsen, 1993) to research, design, build and test an ASR system for the Court. Indeed, I constructed a proof-of-concept prototype of such a system and demonstrated it, live to an audience, with the Chief Magistrate as the user.

After revisiting the Court and imagining the use of the proof-of-concept in the courtroom, I realised that the task of communicating outcomes was more complicated than I had understood it to be. A lot more work was being done than just the magistrate speaking and writing an outcome. Through some preliminary fieldwork at the Court it emerged that the magistrate's act of speaking an outcome was not an event that was self contained but the beginning of a work process distributed in space and time throughout the Court. This prompted a deeper ethnographic investigation of the processes involved in determining and recording outcomes of cases. It emerged that the process involved many different Court workers, each performing detailed work that contributed to the recording and communication of an outcome.

The distributed nature of the work process of the Court meant that a simple ASR dictation system that was situated at the magistrate's bench would not fit with the extensive process that was in place at the Court. A distributed work process built around ASR was not a software project to be taken lightly. Any system that was to be considered for the Court would need to respect the distributed process and work with it, rather than attempt to change it, because the process is a functioning part of the wider system.

Because I wanted to respect the Court's established work process and disrupt it as little as possible, I had to find out how much disruption would be brought about by the introduction and use of ASR software. The next step was to find existing users of ASR software and study them and how they managed to work with the software.

I began two ethnographies of workplaces where ASR software had already been introduced to try to understand the changes that ASR brings. More specifically, I was looking for the changes the ASR software makes on a work place, on established work practices and what, if any, changes the established and modified work practices make to the software. These ethnographies revealed that the introduction of ASR technology to individual users had effects throughout the whole business, even to users who did not work directly with the ASR users.

There are two workplaces I observed ASR software in use: the first was the Hansard department at Parliament House and the second was composed of various offices within the Public Service in general. Since the software used in both locations was basically the same off-the-shelf software, the different realities and perceptions of its use in each location are interesting and revealed what it actually takes to make ASR software usable in a productive environment.

Having explored what made dictation systems usable and useful in real work environments I was able to return to the question of ASR at the Court.

At the Court, through several meetings with the Chief Magistrate, it emerged that he had quite a fantastic view of ASR dictation software and what it could do. He seemed to imagine, though he never put it in these words, that dictation software was a computer-based incarnation of a secretary who could take down dictation and produce a perfect transcript. This is in contrast to the view that ASR is a technology that makes mistakes and requires effort to use. The highly optimistic view of ASR is reflected in science fiction literature as well as the popular press.

1.2 Automatic Speech Recognition in Popular Culture

When investigating ASR, it is common to come across the assumption that it is easy to use because speaking to a person is so easy. To explore why ASR is perceived as easy, this section looks at ASR in popular culture through fiction and the popular science and business press to investigate how ASR is portrayed in non-technical settings and so, portrayed to people not familiar with the technology. Looking, briefly, at how ASR is portrayed shows that it is often presented uncritically, as a somewhat magical technology that is just the same as talking to a human. The reason the uncritical presentation of ASR is useful in this thesis is that many potential users of ASR form their opinions of it through the information presented in newspapers and magazines, and even television and movies.

Automatic speech recognition is one of the holy grails of computing. A computer that can be spoken to, and can speak back, has been imagined since computing began. Even before computing, the idea of inanimate objects that could speak and be spoken to was part of popular culture.

> In fairy tales, characters don't type instructions to a Magic Mirror on a keyboard; they talk to it. (Kaku, 1998)

In Snow White (Grimm and Grimm, 1975) the queen speaks to the Magic Mirror, saying: "Mirror, mirror, on the wall, who is the fairest of them all". The mirror first flatters the queen, saying that she is very beautiful, before announcing that Snow White, the queen's stepdaughter is the most beautiful of all.

In ASR terms, the queen addressed the mirror using *keywords* ("Mirror, mirror on the wall"), she asked a *question* ("who is the fairest of them all") and received a *response*.

The question the queen issued to the mirror was ambiguous. We (the reader) know what "fairest" means in the context of the story, and indeed who "them" means, but why should the mirror? The mirror, enchanted, bewitched or otherwise magical, had some sort of intelligence so that it could understand the queen and know that she meant "most beautiful" when she said "fairest" and which subset of all people alive was meant by "them".

It may be possible to trace the origins of ASR to fairy tales, but most ASR in popular fiction is located in what some have called modern fairy tales, science fiction.

> To have a true Magic Mirror, in essence, involves perfecting artificial intelligence, the most difficult problem of all in computer technology. (Kaku, 1998)

1.2.1   Automatic Speech Recognition in Science Fiction

Not having true automatic speech recognition applications is no impediment to authors of science fiction who frequently use ASR for much the same reason that characters in fairy tales talk to Magic Mirrors – it allows the author to have inanimate objects as characters or it moves the narrative along without having to explain typing or other non-verbal interaction. Because characters in books talk freely to each other, and because in science fiction computer programs can serve as characters, in some science fiction stories talking to computer programs becomes commonplace. In the examples below I have selected some typical examples of characters talking to computer programs that understand them and in some instances talk back. These examples show how different authors treat ASR with varying degrees of transparency. In *1984*, Orwell 2003 introduces the 'speakwrite' as a contrast to the pen-and-ink that the character Winston uses to keep his diary. It is also implied that the speakwrite is the preferred means for doing work because it allows the easy overhearing of other people's work, ensuring that everyone's work is party approved. In some stories, ASR is accepted as being something that works and in others the lack of sophistication in ASR is used to highlight the relative lack of technical sophistication in the world that the author has created.

In Neal Stephenson's *Snow Crash* (1992), people and programs interact inside a world constructed in software that in part models the real world. The software is called "the Street". Inside the Street, the main character of *Snow Crash*, Hiro Protagonist appears as a human, as does the Librarian, a piece of software. How humans represent themselves inside the Street is largely open-ended. Most participants in the Street choose to present themselves as people however this is not strictly necessary. A generic name for a user's representation inside the Street is an "avatar". The quality of one's avatar can be interpreted as an indicator of one's status. In one section Hiro has a long interaction with the Librarian inside Hiro's software-house on the Street. Hiro's house on the Street resembles a traditional Japanese house with tatami mat flooring. Hiro has discovered a new room has been added to his software-house and that a friend of his has gifted him the room, the Librarian and another piece of software called the Library. Initially Hiro and the Librarian engage in some small talk and finally Hiro orders the Librarian to collect some information on one of the shadowy antagonists in the plot:

> " 'Okay, Let's get some work done. Look up every piece of free information in the Library on L. Bob Rife and arrange it in chronological order. The emphasis here is on *free*'.
>
> 'Television and newspapers, yes, sir. One moment, sir,' the Librarian says. He turns and exits on crepe soles."

Hiro is able to converse with the Librarian as if he (the Librarian, who appears to be a "pleasant, fiftyish, silver haired, bearded man with bright blue eyes") is just as intelligent as any human character. The Librarian is able to understand that "free" means non-proprietary, public knowledge and is able to interpret Hiro's request accurately.

Some time passes and the Librarian returns with the information Hiro requested on a *hypercard*.

" 'Your information, sir,' the Librarian says.

Hiro startles and glances up. [...] Like any librarian in Reality, this Librarian can move around without audible foot falls.

'Can you make a little more noise as you walk? I'm easily startled,' Hiro says.

'It is done, sir. My apologies.'

Hiro reaches for the hypercard. The Librarian takes half a step forward and leans towards him. This time, his foot makes a soft noise on the tatami mat, and Hiro can hear the white noise of his trousers sliding over his leg."

The Librarian is not only able to show that he understands information requests but is also able to modify various settings such as the amount of noise that he makes as he walks. This shows that the Librarian, though he acts much like a person, is aware that he is a program. Later, Hiro and the Librarian engage in some discussion on the identity of the Librarian's author and also on how the Librarian's internal structure makes him a "sucker for non sequiturs." but he is still able to converse naturally with Hiro and interact with the artificial world of the Street in the same way that a human's avatar would.

In *Idoru* (Gibson, 1996) ASR is taken to a further extreme where it is used as part of an instantaneous translation system that is part of small "ear-clips" that contain a software program for translating English and Japanese. Not only does the computer understand English and Japanese, it is able to translate between them, in real time, almost flawlessly. In the section quoted below, Chia has traveled from San Francisco to Japan and is talking with Mitsuko. Chia and Mitsuko are members of two different chapters of a fan club for a pop-idol. Chia speaks no Japanese and Mitsuko does not speak enough English. They use the instantaneous translation system so they can talk.

"Now they were both wearing wireless ear-clip headsets. The translation was generally glitch free except when Mitsuko used

> Japanese slang that was too new, or when she inserted English words that she knew but couldn't pronounce.
>
> Chia decided to change the subject. 'What's your brother like? How old is he?'
>
> 'Masahiko is seventeen,' Mitsuko said. 'He is a 'pathological-techno-fetishist-with-social-deficit',' this last all strung together like one word, indicating a concept that taxed the lexicon of the ear-clips. [...]
>
> 'A what?
>
> 'Otaku,' Mitsuko said carefully in Japanese. The translation burped its clumsy word string again.
>
> 'Oh,' Chia said, 'we have those. We even use the same word'[2]"

Idoru is set in a much closer future than Snow Crash and one of the ways that Gibson shows this is that the ear-clips are not perfect. The translation program cannot cope when one speaker uses a word that is outside the lexicon or does not have a direct translation, such as Otaku. Chia and Mitsuko discuss Otakus further and Mitsuko eventually says of Chia's experience of Otaku in America, "I do not think it is the same thing". The ASR in this section is still "generally glitch free" and the times when it makes a mistake are used by Gibson to highlight the differences between the America and Japan of the world he has constructed.

In a scene in *Neuromancer* (Gibson, 1995) ASR is used casually, as if it were an accepted, everyday part of the world. In the scene below, the anti-hero Case is asleep, his head covered by the silk of a sleeping bag, when a voice wakes him:

> "The security package taped to the steel firedoor bleeped twice. 'Entry requested,' it said. 'Subject is cleared per my program.'
>
> 'So open it.' Case pulled the silk from his face and sat up as the door opened."

---

[2]This is not strictly true. Chia understands "Otaku" to mean something closer to the English or American terms "geek" or "nerd".

A character named the Finn enters and finds Case newly awake which is uncomfortable for them both. In this instance, the ASR system is used as another way for Case to be somewhat embarrassed in a slightly awkward social setting. If he had been required to physically answer the door, Case could have been better prepared for the Finn's entrance. Case's casual "so open it" demonstrates the sophistication of the ASR, in that it can recognise and act on such an oblique phrase.

It is unusual for ASR to be portrayed as less than perfect. It is common for speech among humans to be portrayed as misheard, or inaccurate, however to portray ASR as flawed would possibly be making a point that is of little use in fiction. One of the most famous talking computers, HAL, from *2001, A Space Odyssey* (from Arthur C. Clarke's book and Kubrick's film) is flawed, perhaps even insane, but HAL's speech recognition and synthesis is perfect.

In an episode of *Star Trek: The Next Generation* titled *Elementary, Dear Data* (Lane, 1988) Data, an android, enjoys re-enacting Sherlock Holmes mysteries with his friend, Geordi. They play in the "Holodeck" a simulation room aboard the *Enterprise*. Geordi becomes frustrated with Data because Data has read all the mysteries and knows how they end when they are just beginning – there is no chase as Data solves the mystery before it has started. Dr Pulaski says that she does not believe that Data could solve any mysteries as he is just a computer. Geordi suggests that they try solving a new, original, Sherlock Holmes-like mystery, one that Data has not encountered. Data agrees. Geordi asks the computer to synthesise a new mystery in the Holmes style with an opponent who has the ability to defeat Data. The computer asks, "Define parameters of the program.", and Geordi repeats, "Create an adversary capable of defeating Data.". The computer complies and the pair enter the Holodeck to play in the new mystery. Unfortunately, the computer has interpreted the request to create an opponent the equal of Data literally and has constructed a new artificial intelligence in the form of Moriarty. In the course of the story, it becomes apparent what has occurred and the new AI, Moriarty, discovers that he is a synthesis and becomes determined to preserve his existence.

Although it is a small point, in *Elementary, Dear Data*, the plot hinges on the computer taking Geordi's simple request to its most logical extreme. The computer knows that Data is an android, an artificial intelligence, so it constructs a new artificial intelligence that is "capable of defeating Data".

Fiction may be the way that some people first encounter ASR as a concept. Another way to have first contact with ASR is through the popular science and business press.

1.2.2 Automatic Speech Recognition in Non-Fiction, the Popular Science and Business Press

Automatic speech recognition is often presented in the popular (as opposed to academic) press as being a device to save money. Articles are often non-critical and tend to focus on cost savings and benefits of the technology (for example Fox, 2003; Ziff Davis Media Inc., 2004; Macinnes, 2004; Murray, 2004; Anthes, 2004; Reed Business Information Ltd., 2004; Bednarz, 2004; Soto, 2004). ASR frequently does bring benefits to the organsiations that implement it, however there are few articles that deal with the challenges of integrating ASR into an organisation before benefits can be realised.

The article quoted below is typical of that directed towards business people. It presents ASR as a cost saving device, one that can reduce errors and improve productivity.

> "Grocery wholesaler James Hall has cut costs and administration time by investing a six-figure sum in a voice technology system to help staff pick out orders.
>
> [...]
>
> In the first 12 weeks picking errors fell by 90% to 0.01%. In the past four weeks errors have been almost eliminated, with only one error in 60,000 cases.
>
> [...]
>
> The wholesaler expects the investment to pay for itself within 18 months. "The majority of the payback will be generated by reductions in the cost of errors, fewer financial claims, lower

> return handling costs and improved wholesale and retail stock
> accuracy", said Hall." (Gomm, 2004)

The system described has clearly been successful as the return on in-
vestment is quickly realised. However, the article makes no mention of
the difficulties involved in integrating the ASR system into the workplace
or even the time involved to implement it.

An important point that is not noted in the above article, and indeed,
in very few articles, is that not all ASR is the same. ASR systems have
different vocabulary sizes and work in different application domains.

The type of system described above would have quite a small vocabulary. The stock-pickers only have to speak the digits on a barcode, a total
of ten distinct words. A small vocabulary system is more likely to be successful than a large vocabulary one because, as the vocabulary increases
in size, the likelihood of recognition errors increases.

One of the most successful applications of ASR is over the telephone
where it can replace a key-press menu structure. Articles on this sort of
ASR often state that speech is the most natural interface over a phone
(Macinnes, 2004). Dialog interfaces are described further in section 2.3.

These sort of articles often quote the cost savings of using an ASR application to handle customer calls. The cost savings can be very compelling.
In one instance it was stated that the costs of answering each call were
reduced from $3.10 to $0.17 (Macinnes, 2004).

Articles of the "cheerleading" sort may sometimes present a potted history of ASR applications before presenting the latest innovation and associated cost savings. Potted histories tend to gloss over the limitations of
ASR in favour of the benefits (Anthes, 2004).

The other kind of article on ASR is the "perfect future" article that describes a seemingly miraculous application. Increasingly these describe
applications for use in non-desktop applications where an ASR system
is intended to replace or augment existing interaction styles. Such articles
will quote accuracy rates uncritically before going on to describe proposed
applications for future ASR systems. Again, such articles ignore the difficulty in integrating ASR into existing working styles and, in my view,
encourage unrealistic expectations of the benefits of ASR systems.

> "Speech recognition technology with a claimed accuracy rate of
> 95 per cent has been developed by researchers at Toshiba. The
> middleware, which also features a text-to-speech (TTS) engine,
> is available from Q3 this year and supports nine languages.
>
> Toshiba said it expects ASR to become standard technology for
> in-car navigation and telematics, home multimedia gateways,
> and more powerful handheld devices. A key application for
> the TTS engine is read-back of emails." (Reed Business Infor-
> mation Ltd., 2004)

In the above quote the intermingling of text-to-speech and speech-to-
text (that is, ASR), blurs the distinctions between the two technologies,
making it difficult for the reader to understand them as separate problems.

Articles in the business press can often present ASR as part of a more
complex system that involves many technologies working together. In
some cases, a computer system may be identified as ASR when it is only
part of a more complex software package. In the quote below such a system is presented as understanding 1.7 million spoken street names as well
as being able to be a computerised concierge:

> " 'If you ask for directions to an Italian restaurant, it can find
> the restaurant, display the review on a screen or read the re-
> view back to you while you drive', said Alister Rennie, vice
> president of sales and marketing for IBM Pervasive Comput-
> ing." (Murray, 2004; Honda Motor Company Ltd., 2004)

The system described above would need to have access to an ASR component, a restaurant review component, a speech synthesis component,
and a navigation component, however the article describes the system as
a "hands free and natural sounding in-vehicle speech recognition system".
Articles such as these present unrealistic views of what ASR is and is not
(though it can sometimes seem that ASR is the answer to all problems).
In particular, articles in the popular press skew the opinions of decision
makers towards believing that ASR is simple, easy and "natural" and can
be used in any situation where people *might* speak to a computer.

Automatic speech recognition has also been used, in various forms, in court rooms in the United States for producing transcripts live in court. It has even been called "the court technology for the 21st century" (Polansky, 1997). Because ASR can't accurately recognise the speech of many people in a court environment, live transcription is achieved by a single ASR user sitting in court and repeating every word said by all of the parties in the court into a computer running ASR dictation software. Since repeating every word spoken could be very annoying and disruptive to the court proceedings, most "voice writers" or stenomask operators dictate into a mask that resembles a gas mask. Voice writers typically use off-the-shelf dictation software to perform their job, though more recently it has been possible to obtain specific software packages or add-ons for court transcription.

Voice writers are the cause of some controversy in the (North American) National Court Reporters Association which primarily has stenotypists as members (Buckley, 2002; Pennington, 2002). Voice writers are variously seen as a threat to the stenotypists jobs or as not being able to produce transcripts of sufficient quality to compete with stenotypists (Huber, 2000; JCR Online, 2000, 2001; Meadors et al., 2001).

1.2.3 Summary of Automatic Speech Recognition in the Popular Press

Automatic speech recognition software is presented in the popular press as akin to magic—it is usually a perfectly smoothly functioning technology, which as Arthur C. Clarke said, is indistinguishable from magic (Clarke, 1974). In science fiction it is usually part of the background of the invented world. The sophistication of the ASR is used to subtly show the degree of sophistication of the technology in that world.

As discussed in the previous section, section 1.2.2, in the popular science press and in business journals, the use of ASR is often reported positively with little regard for the difficulties in implementing and using an ASR system. Where ASR has the potential, or at least the perceived potential, to displace the skilled work of people, those people debate its merits.

None of these views align with the real experience of people who use ASR in their daily work.

1.3 Research Questions

The focus of this thesis changed from the ACT Magistrates Court specifically to workplaces where dictation systems are used, and might be used, with the Court as one example of such a workplace.

The question I asked when I began this thesis was: What form could ASR take at the Court? Through a process of research evolution and discovery, the questions my research now asks, and answers, are:

- How is ASR used and made usable and useful in the workplace now?

- How could ASR be used and made usable and useful in workplaces in the future?

The workplaces that this thesis looks at now are, as described above, various instances of the work of public servants and the Hansard department at Parliament House. The workplace where ASR could be used is the ACT Magistrates Court.

This thesis still answers the question, *What form could ASR take at the ACT Magistrates Court*, but treats the Court as an instance of a workplace where ASR might be used in the future. This thesis addresses the wider issue of dictation software in the workplace and how such software is made usable through the efforts of users and other stakeholders.

1.4 Thesis Statement

My thesis is that talking to a computer is difficult. To be more precise, talking in such a way to be understood by an ASR system is different than talking to be understood by a person because an ASR system does not have the human capability to listen for meaning. Large vocabulary dictation style ASR systems are hard to use in the workplace. The difficultly in using dictation systems comes not from poor recognition accuracy but

from: (1) different expectations that people have of what such systems
can do; and (2) the integration of ASR systems with work processes and
work organisation. Designing ASR dictation systems is a matter of helping (potential) users have realistic expectations of the system and making
the integration of dictation systems with the technical and non-technical
elements that make up work as simple as possible.

This thesis shows how the properties described above contribute to
the successful use of ASR dictation systems in the workplace and then
describes how those properties can be used to lead to the design of future
dictation systems.

The user's objective is not the use of ASR. The objective is always to
perform a task or series of tasks. In the work situations described in this
thesis, ASR systems act as *middleware* by acting as an intermediary between a user and the tools that they use to perform their tasks.

Middleware is software that acts as a bridge, usually between two
other pieces of software.

1.5   Thesis Plan

This chapter has described the evolution of the research question and the
research design from a software engineering approach to a qualitative
study of three workplaces. It has also shown how ASR is presented in
the popular press. Finally, this chapter has stated the thesis, that talking
to a computer is difficult, and related that thesis to the main theme of this
research, that usability is contingent and co-produced through the interaction of many elements both human and non-human, social and technical.

In the next chapter, I examine how ASR is portrayed in academic usability literature. The recent literature on the usability of ASR applications
focuses on recognition accuracy as a measure of the usability of ASR, with
little regard to the integration of such applications into the workplace. This
is in contrast to user-centred design principles where the user's task has
primacy. It is in the space of designing ASR systems for the support of
users' tasks that this thesis is situated.

In chapter 3, the methods used to conduct the qualitative investigation and analysis used in this thesis are introduced and their use explained.

Chapter 4 contains two ethnographic accounts of ASR dictation users. The first is an account of the work of the Hansard editors; the second an account of the work of ASR application users in the Public Service who suffer from varying kinds of workplace overuse injuries. These two accounts are briefly compared before being analysed in chapter 6.

Chapter 5 is an account of the work of recording sentencing outcomes in the ACT Magistrates Court. It is shown that the recording of outcomes is a process distributed in space and time throughout the Court.

Chapter 6 presents an analysis of the work practices of the Hansard editors and the Public Service ASR application users. Using concepts introduced in chapter 3, a trajectory of use of dictation systems in the workplace is described. The trajectory encompasses nine properties that have been identified as essential for the successful use of dictation systems in the workplace.

Chapter 7 analyses the work of the Court with particular regard to the opportunities to introduce ASR into the process of recording outcomes. Using a scenario-based approach, the difficulties in introducing ASR to the Court are described. A design for a re-imagined ASR system for the Court is also presented.

Chapter 8 is a reflection on the methodologies used to analyse the ethnographies and design the novel system for the Court.

Chapter 9 concludes the thesis and suggest directions for future work that can continue the themes of this thesis.

CHAPTER 2

A Review of the Literature on the Use and Design of Automatic Speech Recognition

In this chapter I review the literature relevant to studies of the usability of large vocabulary automatic speech recognition (ASR) systems. The usability of speech recognition software is often considered to be related to recognition accuracy and much of the literature reviewed here takes that viewpoint. The first section of this chapter introduces speech recognition as a general field of research, a field of interest to linguists, designers, psychologists and engineers. Starting with the first studies into the potential usefulness of an ASR system (Gould, 1978; Gould et al., 1983) I will show that research into the use of large vocabulary speech recognition systems maintains a focus on recognition accuracy despite other work showing that the entire system must be taken into account when analysing the usefulness of an ASR system.

Large vocabulary speech recognition is not the only kind of speech recognition and not the only kind of interest to interaction designers. Small vocabulary speech recognition systems, for example dialog-driven systems (for example Yankelovich, 1996), are being used over the telephone in situations that may have previously used a menu-driven interface of the "press 1 for billing inquiries, press 2 for new accounts" sort. A lot of usability work is done with dialog-driven speech recognition interfaces (for example Rosenfeld et al., 2001; Weinschenk and Barker, 2000; Shriver and Rosenfeld, 2002). Usability of dialog-driven interfaces has a lot in common with screen-based usability being concerned with intelligibility and navigation.

2.1 Motivation for looking at Speech Recognition

Speech recognition is assumed to be a more "natural" way of interacting with a computer because speech is one way in which people interact with each other. Other "natural" interfaces are gesture recognition and handwriting recognition. Natural interfaces are often compared to keyboard and mouse interfaces with natural interfaces assumed to be the necessary direction for computing to be taken "off the desktop" (Abowd and Mynatt, 2000). The largest problem with interfaces that use a recognition paradigm is that they are error-prone in their interaction. This is not to say that it is easier for users of speech recognition systems to make mistakes with such systems but that the systems themselves make mistakes. For the purposes of this thesis, these mistakes are akin to transcription errors made when a word or phrase is miss-heard. For example, the word "Australians" may be incorrectly recognised as "astray aliens". Not all speech recognition errors are as obvious to humans as Australians/astray aliens and it can be difficult at times to determine how an error occurred. The reasons for this propensity for error is not the topic of this thesis; it is enough to know that they occur and that the research community are attempting to minimise the errors through various means. Until the propensity for error is overcome[1] it is a feature of recognition interfaces. The "disobedience" (Snape et al. (1997), cited by Read et al. (2002)) of recognition interfaces makes them harder to use and harder to understand in use. A focus on the

---

[1]If, indeed it can ever be overcome as even humans make recognition errors (Abowd and Mynatt, 2000, pg. 34)

disobedience or propensity for error is a defining characteristic of most research on the usability and usefulness of recognition interfaces. Still, the undeniable fact that recognition interfaces make errors which makes them hard to use does not stop researchers treating recognition interfaces as "natural".

Articles saying that speech recognition is an "interface revolution" are prevalent in the technical press (for example Clark, 2001) and in academic literature (for example Cohen and Oviatt, 1995). Both types of articles focus on the increase in handheld devices and the difficulty in using such devices through the typical interaction means of keyboard and mouse. Some even mention stylus-based interaction as equally problematic because of the combination of small screens and the difficulty in building complex interfaces that can be manipulated on a small screen with a stylus (Clark, 2001). The optimism of such articles leads them to only ever briefly mention the difficulties in building speech recognition devices and the difficulties of using them. That such articles are somewhat biased is natural given that they are written from a general point of view.

In the academic literature a (usually) more balanced view is taken when comparing speech recognition with other methods. Researchers in the area of speech recognition are often optimistic that the use of speech recognition will increase when the recognition accuracy improves so that a recognition system is indistinguishable from a human listener. Currently, given an ideal environment, computer speech recognition approaches the accuracy of human speech recognition, though the computer will typically have an error rate that is an order of magnitude greater than the human and this difference increases as the quality of the audio signal deteriorates (Lippmann, 1997). Cohen and Oviatt (1995) identified five situations where speech recognition could be a preferred interface: "when the user's hands and eyes are busy; when a limited keyboard and screen is available; when the user is disabled; when pronunciation is the subject matter of computer use and; when natural language interaction is preferred." In my research I have looked at situations where users are disabled (which are described by Cohen and Oviatt) and situations where natural speech is the input to a work process[2], a situation that Cohen and Oviatt did not describe. Cohen and Oviatt say that successful use of speech recognition systems often happens when there is no adequate alternative to use when interacting with a computer and they note that there are questions as to whether people will use speech recognition if an alternative means exists. Significantly for my research, they also state, "it is not obvious why people should want to speak to their computers in performing many tasks—in particular, their daily office work". Despite this small reservation, Cohen and Oviatt are speech recognition optimists who are looking to advance the cause of speech recognition by gradually improving recognition accuracy so that speech recognition systems may be usable in their five identified situations.

Cohen and Oviatt's view of the design of speech recognition technology is technology-driven, a view that is specifically described by Danis and Karat (1995). Technology-driven design tells technologists, briefly, to build new speech recognition systems that are "grounded in real work contexts", and put them out into the real world for use even though the technology is not yet perfected. When the technology is used in the real world, issues for designers will become apparent and those issues can be fixed in future iterations of the technology. This approach, according to Danis and Karat, is tempered by a commitment to user-centered design that prevents unusable products being imposed on the general public. However, studies of speech recognition systems, particularly large-vocabulary speaker-dependant speech recognition systems that are the focus of this thesis, have remained in the laboratory with a focus on recognition accuracy as being the determiner of the usability of such systems in contrast to smaller vocabulary systems, eg dialog systems, where good recognition accuracy and an orientation to the task at hand has seen those systems become more widely used in the real world. If the optimism of Cohen and Oviatt is to come to fruition for large vocabulary speech recognition systems then studies of such systems must go into the real world and cannot remain in the laboratory. Danis and Karat's program of technology-driven design does not seem to have been followed for large-

---

[2]That is, the ACT Magistrates Court and Hansard.

vocabulary systems which have been seen by researchers as not yet ready for the real-world, despite being sold as shrink-wrapped software! Studies of such systems have remained in the laboratory, despite it being possible to study such systems in the real work situations. The next section describes laboratory studies of dictation systems.

Speech recognition pessimists (Shneiderman, 2000; James et al., 2002) point out that using speech recognition results in a higher cognitive load than using non-speech input methods. Even in situations where speech recognition might be considered useful, for example in the cockpit of a fighter jet when the pilot's hands and eyes are elsewhere, using speech recognition has been shown to be problematic. Still, even speech recognition pessimists do not say that the technology should not be used, they are merely a lot more circumspect about its value than the optimists are.

2.2 Usability Studies of Dictation Systems

Usability studies of speech recognition systems have tended to focus on the rates of recognition errors in test conditions of the systems as a quantifiable measure of usability. More elaborate studies have used error rates as a starting point and classified the sorts of errors that occur in the hope of being able to design around particular classes of error.

It can be argued that John Gould started the field of usability research in speech recognition systems when he studied how experts dictated business letters (Gould, 1978). Gould tested business executives who were experienced dictators and novice dictators. He had a pool of subjects available because he worked for the IBM Research Center and the studies were performed in the 1970s when it was still common practice for executives to dictate correspondence. Gould tested eight subjects on four modalities: writing, invisible writing[3], dictating, and speaking[4]. There were two tasks that the subjects completed in each modality. In the first task they had to

---

[3]i.e. writing without being able to see the paper.

[4]The difference between dictating and speaking being that in *speaking* the subject was told to assume the recipient would listen to the letter but in *dictating* the subject needed to give typing instructions as well as compose the letter.

write a simple business letter and in the second they had to write a "complex letter" on a particular subject, for example, applying for a job or the significance of the United States Bicentennial. Gould videotaped the participants and had them rate their own letters and the letters of the other participants for quality.

Gould found that learning to dictate did not take a particularly long time with novice dictators achieving similar dictation speeds to the experienced subjects by the end of the testing. He also found that dictation was faster than writing, though not as fast as some figures had indicated. The tests found no quality difference between written and dictated letters and also found that when all the letters were typewritten, participants could not reliably pick which letters had been dictated and which had been written. Speaking was found to be more "natural" (Gould's quotes) than dictating with the experienced dictators being significantly faster than the novices. Writing and invisible writing, surprisingly, had about the same composition time and quality ratings, leading Gould to suppose that having the previously composed text visible was not as essential as it might intuitively seem. The experiment lead Gould to conclude that output modality was not related to performance and that skill at composition was the determining factor in high-quality letters.

Gould followed the "How Experts Dictate" experiment with an experiment in composing letters on a "simulated listening typewriter" (Gould et al., 1983). The listening typewriter experiment was intended to test whether an imperfect listening typewriter would be useful for composing letters. To test this, Gould et al. simulated a listening typewriter with a limited vocabulary of 1000 or 5000 words and an unlimited vocabulary version. Gould et al. compared the listening typewriter with handwriting and traditional dictation to tape. The listening typewriter had two different "implementations": isolated word speech and connected word speech. In isolated word speech the user had to pause between words and in connected word speech the user was able to speak faster but in a still unnatural style. They hypothesised that:

- The participants could write letters faster by handwriting than by using the listening typewriter because of their greater familiarity with writing;

- Connected word speech would be faster than isolated word speech;

- The quality of the letters from each method would be the same, and;

- It would take longer to proofread letters composed on the listening typewriter than handwritten ones because of the extremely limited ability of users to edit while they dictated to the simulation.

The simulation was particularly ingenious. Users were not told that the listening typewriter was a simulation but believed it to be working technology. In reality the user was speaking to a secretary who was using a computer. The secretary would type the word spoken by the user and hit "return". The computer would check if the word was in the allowed dictionary of 1000 or 5000 words and if it was, it would display the word for the user. If the word was not in the dictionary, the user would see "XXXX". Another "error" that was built into the system was for homophones[5] to be displayed as the most common occurrence regardless of context or how the word was typed by the secretary. The simulated system also had quite a complex set of formatting and navigation commands[6].

In the first experiment with the listening typewriter, participants had to compose a letter to convince another person of something. For example, they had to apply for a job, request a loan or recommend that their office be relocated. Participants generally preferred using the listening typewriter to handwriting letters.

In a second experiment with eight subjects from the IBM organisation, participants used the listening typewriter in its various modes and dictated to a dictating machine and a secretary who took shorthand. Composition time was fastest when dictating to the secretary but was only significantly faster than the isolated word condition for the listening typewriter.

---

[5]A homophone is a word that is pronounced the same as another word but spelt differently. An example is "which" and "witch".

[6]Of particular amusement to me was the command for having made a mistake: "Nuts"! Other simulated commands were more formal.

Gould et al. note that participants found the simulation to be "extremely compelling" and when told, in the middle of the experiment about the workings of the simulation, some even argued as to why they believed that the system was "real".

Gould et al. conclude by saying that they believed that people would be able to use true (not simulated) listening typewriters to write letters. They go so far as to say that they believed there would be productivity benefits because no typing would be required and there would be faster turnaround of dictated letters. Participants told the experimenters that the vocabulary size was more important than the speech mode and the results indicated that a 5000 word vocabulary isolated speech system would be usable, though the authors note that the participants were not particularly enthused by the system.

Significantly, they found that the subjects noticed a "major difference" in accuracy rates of 91 and 100 percent, a finding that was validated later by Van Buskirk and LaLomia's work on the "just noticeable difference" of speech recognition accuracy (1995). Van Buskirk and LaLomia found that the just noticeable difference for accuracy of transcribed text was between 5% and 10%, meaning that most people cannot tell the difference between a 95% correct transcript and a 90% correct transcript. This also means that the next noticeable step for speech recognition from currently stated accuracies of about 95% is 100% accuracy.

In a recommendation that can be seen to have influenced research that followed the listening typewriter simulations, the authors said that they did not think that an improved editing system would have allowed the subjects to compose faster but might have improved their feelings toward the system. Current editing and correction systems can act as bottlenecks in the use of speech recognition systems (Feng, 2002). This finding corresponds with Moore (2004)'s work on modelling data entry rates for ASR and other systems. Moore described a model that predicted data throughput as a function of data entry rate in words-per-minute, error rate as a percentage and the time to correct each error. According to Moore, speaker dependant large vocabulary ASR, that is typical desktop dictation systems, has a high entry rate but a very low throughput rate because of

ASR's high error rate and long error-correction time.

The findings of Gould on what users wanted from dictation systems was influential in the direction of future research in speech recognition systems.

At the same time, limited vocabulary speech recognition was examined at two different work locations by Rollins et al. (1983). In examining the use of speech recognition in two different workplaces, the authors said:

> The data suggests that success of a speech recognition device is related to variables describing the entire system, e.g., complexity of host software, or task complexity. The variables more closely relating to speech input, e.g., microphone type, appear to be less of a problem.

Gould noted the importance of error correction and recognition accuracy. Rollins et al. noted that it was the factors associated with the complexity of work that were most influential on the successful use of ASR in the workplace. It is this second point that the remainder of this chapter addresses.

MedSpeak (Lai and Vergo, 1997) is a dictation system that was developed by two researchers at IBM Research, Jennifer Lai and John Vergo, for radiologists. The radiologists had to create reports on the x-rays that they viewed and MedSpeak used ASR as a way for radiologists to create, edit and manage reports faster than what was possible using the previous system of tape recordings sent to a typing pool.

The workflow for the radiologists using the typing pool system was to dictate the information on the requisition form for the report then dictate the report while looking at the "film" (i.e. an x-ray). The radiologists dictated very quickly and typically produced about 100 reports a day. Trained medical transcriptionists produced typed reports which were sent back to the radiologists for a signature. The problem with this system was the reports could sometimes take more than 24 hours to return to the radiologist by which time they could not remember their original report. They could not check the report for accuracy of transcription, having forgotten the x-rays they had seen, and could only check for mistakes made by the typing pool.

MedSpeak was developed to circumvent the typing pool and allow radiologists to dictate reports and then view the typed report immediately for correction and signing. MedSpeak was developed with a radiology-specific vocabulary and language model to allow the software to recognise the specialist vocabulary of the radiologists.

The main advantage of MedSpeak was that report turnaround time was decreased. The authors state that turnaround time was reduced from up to 50 hours to "a couple of minutes".

Interestingly, the authors also reported on the aspects of the system that were not liked by the radiologists. One of these aspects was the use of computers that the system required because some of the radiologists were not especially familiar with using a keyboard and mouse. MedSpeak also led to a change in the role the radiologists played. A radiologist was quoted as saying, "when I put my hands on the keyboard I am doing an administrative task and no longer functioning as a physician".

MedSpeak was designed for high recognition accuracy as well as to fit with the radiologists' work. Where the pre-MedSpeak work process was long and drawn out, with MedSpeak reports were available for checking as soon as they were dictated leading to better medical outcomes. The success of MedSpeak was not due to high recognition accuracy, though that was undoubtedly important but rather a good fit with the task at hand. Correction of recognition errors was important in MedSpeak but it would seem that the true measure of the product's success was its integration into the radiologists work.

Antiles et al. (2004) describe the successful implementation of an ASR system for radiologists that took place two years after a failed attempt. Training for users and excellent support were the keys to the success of the implementation.

Dictation systems have also been used in other areas. StoryWriter (Danis et al., 1994) was developed for journalists who had some form of occupational overuse injury (also commonly known as repetitive strain injury, or RSI). StoryWriter allowed the injured journalists to return to work by letting them overcome their injury. The software itself underwent testing with the journalists in the workplace before their comments were incorporated into a revision. The injured journalists liked StoryWriter because it allowed them to return to work, despite being injured. Being 1994, the StoryWriter interface was much less sophisticated than what is available today. It can be assumed that the recognition accuracy was also less than what is available today in off-the-shelf software. Nonetheless, StoryWriter was valuable to the injured journalists.

In the early studies of ASR software it was recognised that integration of the software with the task at hand was important. Inventory storage and baggage handling (Rollins et al., 1983), writing short stories for a newspaper (Danis et al., 1994) and dictating short reports on x-ray films (Lai and Vergo, 1997) are, more or less, narrowly defined tasks. In contrast the conclusions of Gould (1978) and Gould et al. (1983) that error correction is of primary importance for users of large-vocabulary dictation systems seems to have influenced more recent research on such systems.

Three papers (Karat et al., 2000, 1999; Halverson et al., 1999) address the dislike users had for recognition errors and difficulties in error correction and codified the kinds of errors that can occur in ASR systems. Karat et al. (1999) report on usability tests conducted using commercial, off-the-shelf, ASR software. They ran two parallel studies. The first study involved novice users who received limited exposure to the ASR software. The second study involved the authors themselves as long-term users of the same software packages.

Short term users were given some time to train the system to their voice, a task which took between 30 minutes and 1.5 hours, and were then give 40 minutes training time to familiarise themselves with the system. The long term users used each of three ASR products for 10 sessions, each about one hour long. The long term users completed actual work with the systems and after "at least 20 sessions" attempted the same tasks as the short term users.

The tasks were a transcription task and a composition task with each task being completed with ASR and keyboard-and-mouse interaction. Average time for the ASR tasks was 8.74 minutes and for the keyboard and mouse tasks was 3.64 minutes. The long term users completed the same tasks with ASR in 3.10 minutes. The authors then focus on a detailed analysis of the sorts of errors made by the users because, as they state, the overall goal of ASR systems is to "approach the accuracy assumed for users' typing". Two types of errors are identified:

- Redictation failures where the user is attempting to correct an already wrong word and the new dictation is also misrecognised; and

- Cascading failures where a command that is used is misrecognised and must then become part of the correction process.

Each of the ASR systems had a different correction strategy that was recommended for users, with IBM's VoiceType system recommending a two-pass strategy where the user first dictates and then moves on to correction, and Dragon Dictate encouraging an in-line correction strategy. Significantly, the authors say, "to a large extent these strategies were encouraged to have user behavior correspond to system designs and not because of a user driven reason". All of the ASR systems had support for both in-line and second-pass correction strategies. Many of the short-term users used an in-line correction strategy. The authors make the leap from the short-term users' preferred correction strategy to making the recommendation that software support is required for "knowing when a misrecognition has occurred". An example of this support, described by Feng and Sears (2004), uses recognition confidence scores to assist users in making corrections, however the uses of confidence scores to assist in correcting errors has not yet proven successful (Suhm et al., 2001).

Other recommendations to arise from their work are that keyboard-and-mouse interaction with the software be maintained, especially during correction. Obviously this is impractical for injured users.

Feng (Feng, 2002; Feng and Sears, 2004) argues, based on Karat et al's research, that navigation by voice in ASR packages should be simplified from commands that are variable, such as "move back N lines" (where N is a whole number) to simpler "short, fixed" commands such as "move back a line" with the argument that variable commands are harder to remember for users and harder to recognise for computers.

The work on using dictation systems seems to have largely ignored the earlier work on the importance of factors external to the ASR software, e.g., the task at hand, in favour of a focus on the mechanics of ASR and error correction. Error correction is a problem for dictation systems, though focusing on it to the exclusion of wider factors relegates ASR to being a laboratory technology until the errorful nature of computer speech recognition is overcome.

A technique known as multimodal fusion aims to reduce the occurrence of recognition errors by combining multiple modes of interaction to "cancel out" any errors that occur in one mode with correct information gained from a different mode (hence multimodal fusion). Others have suggested that multimodal interaction could aid users in correcting recognition errors (Suhm et al., 2001). Combining two recognition interfaces is not a new idea, the canonical example being "Put that there" (Bolt, 1980) which used speech and gesture as an interface for a variety of tasks. The different modes in multimodal fusion are typically ASR and some form of gesture recognition, including handwriting, though the modes can include "speech, pen, touch, hand gestures, eye gaze, and head and body movements" (Oviatt, 1999b). (see Oviatt (1999a); Oviatt et al. (2003) for examples of research in multi-modal interaction). While multimodal fusion does advance the cause of ASR accuracy, it does so without taking into account wider system factors. Multimodal interaction would also require a significant reorganisation of the way work is done, speech and gesture being very different from keyboard and mouse interaction, though multimodal interaction is not proposed for daily office tasks.

The focus on technical ASR aspects of dictation systems is in contrast to dialog-driven interfaces where a focus on task is at the fore.

2.3 Dialog-driven Interfaces

In addition to ASR research on dictation systems, another area of research for ASR is dialog systems that are typically used over the telephone (Halstead-Nussloch, 1989) in customer-service applications. ASR for dialog applications has much in common with other speech recognition applications and several key differences. The differences all stem from the directed nature of dialog interfaces. Dialog interfaces can be directed because of the explicit turn-taking model (Yankelovich, 1996) they use. The turn taking model has a several main follow-on effects: a smaller vocabulary and smaller perplexity[7] and a lower error rate. Because of the smaller vocabulary that many dialog interfaces use, they can use a much broader voice model, allowing a dialog interface to be used without the training and enrollment procedures that dictation interfaces require[8].

At the same time, the more directed nature of dialog-driven systems forces the designers to consider the task that people will be performing when the system is used. By making the user's task important, designing dialog-driven systems becomes much easier by allowing the designer to make use of "traditional" human-computer interaction techniques and priorities and to ignore "variables more closely related to speech input" (Rollins et al., 1983). In doing so, a designer is able to treat ASR as stable system software and to focus explicitly on what can be seen as typical user interface goals (Vanhoucke et al., 2001), for example navigation and findability and less on the ASR-specific problems frequently seen in large vocabulary speech recognition work.

By being able to treat ASR as stable system software, designers are able to create tool-kit level systems for the construction of dialog-driven systems. The work that has been done on extensible dialog-driven ASR interfaces (for example Rosenfeld et al., 2000a, 2001; Shriver and Rosenfeld, 2002; Rosenfeld et al., 2000b; Shriver et al., 2001) would not be possible if dialog interfaces were as dependent on ASR-specific variables as dictation systems can be.

---

[7]Perplexity is the measure of how many words in the vocabulary are allowed to follow other words.

[8]Many ASR dictation interfaces require a user to "enroll" in the system, or "train" the system, in order for the system to more accurately recognise the users voice. Typically this involves an enrollment stage where the user reads predetermined sentences in order for the system to more closely align its voice model with the users voice. The voice model contains a representation of what the system expects the user to sound like.

2.4 The Importance of Supporting the Task

In the literature on dialog-driven systems (for example Yankelovich, 1996; Rosenfeld et al., 2000a), the researchers build support for the task they imagine their users are going to perform into the systems they build. Because the tasks that people perform with dialog-driven systems are small and self-contained, building support for such tasks into dialog-driven systems is challenging but not impossible. In the same way, the designers of dictation systems build their systems with reference to an imagined task that *their* users will perform.

Goette (2000) studied users of ASR systems who were disabled. Many of the users were quadriplegics and two were blind. Half of the users were successfully using ASR systems and half had failed. The keys to successfully using ASR systems were found to be realistic expectations on behalf of the user and employer; a correct fit between the user's task and the capabilities of the system; adequate and appropriate training; and, trialling the system in the environment in which it would be used. All four keys were related to the fit between the task and the software's capabilities.

2.5 Recognising "Natural" Speech

In both dialog-driven systems and dictation systems the person speaking is aware that their speech is being recognised, allowing them to tailor their speech, and other variables to an extent, to the recogniser. In other speech recognition, the person speaking is either unaware that their speech is being recognised by a computer or they are aware but they are not speaking *for* recognition. These kinds of speech recognition are described in the next section.

ASR has another use where the person speaking need not be aware that they are talking to, or for, a computer. Systems of this sort are typically used in two situations. First, they may be used for broadcast speech recognition, that is recognition of speech from television and radio programs. Typically, broadcast speech recognition is done for archival purposes or to create captions "on the fly". Second, they may be used to capture spontaneous speech from people in meetings or other social situations where it may be advantageous to have a record of what was said.

2.5.1   Broadcast Automatic Speech Recognition

Transcription and captioning of broadcast audio is not a simple problem. In addition to speech, it is desirable that other sounds are captured. Transcription and captioning are slightly different problems though both must deal with the "naturalness" of speech and with the high variability of words that may occur in any particular segment. It is the robustness of such broadcast ASR systems, both for transcription and captioning, that are interesting for this thesis (see Chapter 7).

Broadcast ASR systems for transcription can afford to work more slowly than real-time, often in the order of 130-300 times slower (for example Ligget and Fisher, 1998) because they make multiple passes over the speech input signal to increase the recognition accuracy. Such systems are often not very accurate when compared with dictation systems because dictation systems are trained to a single user's speech and broadcast systems use general voice models. Creating models to use for broadcast recognition (Ando et al., 1998) is an important area of research because the words used in television news programs are typically different from those used in other situations. News programs would necessarily feature certain people's names and other proper nouns more frequently than conversational speech between friends.

Another issue for broadcast ASR systems is that of "data partitioning". Data partitioning is the separation of the continuous audio stream into speech and non-speech sounds. Further separation is also possible within the speech segments by gender and within the non-speech segments by bandwidth for example (Gauvain et al., 1998).

Recognising speech in real-time for captioning is more problematic because hundreds of passes over the signal are not possible. Consequently, real-time ASR is less accurate than multi-pass ASR with word error rates of between 20% and 65% reported in the literature (Wactlar et al. (1998) cited in Ahmer and King (1998)). Captioning systems (Cook et al., 1998) must deal with the same sort of signals as transcription systems, signals that include non-speech sounds including music, speech over music and speech over other speech.

Captioning of television programs is not as simple as providing a direct transcript of the spoken words in the program. The caption must appear when the spoken words are uttered and must disappear when the utterance finishes. Other captions can include sound effects (e.g., "bang!") or music (e.g., "guitar riff"). In Robert-Ribes (1998) the problems in using ASR to automatically caption television programs are outlined. The major considerations in captioning are that the words in the caption are as close as possible to the words spoken while keeping the total number of words on screen below the limit of readability which is about 3 words per second for adults. The author says that any use of ASR for captioning would require human intervention after the automatic process for "manual verification and adaptation" of the captions. This could be a long process with a lot of speech in television programs being corrupted in some way by background noise and with the need for many captions, up to 45% in the study, to include non-speech information such as who is speaking (e.g., John: who's that?) or music or sound-effects. The timing of captions is also difficult with the presentation of captions depending on the timing of speech but also the timing of cuts and other video information. The authors say that correctly generating automatic captions may require an artificial-intelligence system that can watch for cuts in the video. The creation of non-verbatim captions, usually to save time, is also a difficult problem for ASR and would require a speech understanding mechanism. Ahmer and King (1998) encounter similar problems when attempting to create a corpus for use in the creation of new broadcast ASR systems, finding that 75% of speech had been transcribed verbatim and 6% had not been transcribed at all because of information contained in the video image.

2.5.2 Using Recognised "Natural" Speech

Transcripts of recognised natural speech are currently errorful however this does not mean that they are unusable. Researchers at AT&T devel-

oped the Spoken Content-ased Audio Navigation system to turn recordings of broadcast news (Whittaker et al., 1999) and voicemail (Whittaker
et al., 2002) into errorful transcripts. The researchers reasoned that search-
ing audio archives was time consuming and that being able to search
through transcripts would be faster. Multi-pass speech recognition was
used to turn the archived broadcast news and voicemail into errorful transcripts. The transcripts were then used as an interface to the underlying
audio recording. Though the transcripts contained errors they were often accurate enough to allow a user of SCAN to determine if the audio
contained relevant information. Users were able to skim through a large
number of transcripts much faster than they were able to listen to them
and when they found a potentially relevant transcript they could listen
only to it. The errorful transcripts turned unusable speech recordings into
something useful.

Whittaker et al. SCAN is used, and examined in more detail, in section
7.3 as part of the inspiration for re-imagining ASR for use in the Magistrates Court.

2.6   Summary

The purpose of this chapter has been to introduce the current literature on
the use and design of ASR systems. The literature surveyed here illustrates
the gaps in the existing knowledge. A great deal is known about how to
build ASR systems but not a lot is known about how those systems are
used in the workplace. This thesis argues that the design of useful and
usable ASR systems is grounded in an understanding of the real work
that is done using such systems. In order to investigate the use of ASR
systems in the workplace, I undertook fieldwork in a variety of locations.
Analysing fieldwork in order to inspire design requires a variety of tools
which are introduced in the next chapter.

CHAPTER 3

Tools to Think With: Methods Used in this Research

In this chapter I describe the methods and strategies I have used to research and analyse my case studies and to move from those case studies to a design. A variety of the methods are presented in this chapter, from various fields. I have used all of these methods as "tools to think with" to make clear what I have seen in conducting my research and move from that research towards a possible design for an automatic speech recognition (ASR) system.

The methods described in this chapter are different to those typically employed in the design of ASR systems. Where the design approach used in the design of ASR systems is described it is grounded in engineering and the design of machines and software rather than in interface design and an understanding of people. In this thesis the methods described in this chapter are employed to understand how people use ASR systems and what their requirements are for future ASR systems.

In current ASR systems literature work begins, or is at least described as beginning, at a point after which it seem that detailed user requirements have been gathered. For example, Goronzy and Beringer (2005) describe work on a rapid prototyping system for multimodal interaction but do not discuss how user research is incorporated into the process embodied in the tool. Nanavati and Rajput (2005), in describing dialogue call-flows on pervasive devices[1], equate usability with the number of questions the system asks though they do not say how they are able to make this connection. Similarly, Ward et al. (2005) say that they "build on previous attempts to relate usability to a systems technical properties" and though they have an elegant model to illustrate this relationship there is no work with real users to verify it.

Some work in ASR systems considers the people who use the systems. For example, Oria and Koskinen (2002) and Wilkie et al. (2002) describe usability tests of ASR systems.

However the methods described in this chapter are employed at an earlier point in the process of designing a system. The social methods described in this chapter start with a consideration and close examination of people and the situations in which they work *before any work is done on a new system.*

Being exposed to the work of researchers who have been responsible for the introduction of social methods to computer science and computer systems design gave me permission to go "into the field" to conduct my research. The use of social methods in computer systems research has a fairly long history (for example Suchman, 1987; Hughes et al., 1992b; Blomberg et al., 1993; Rogers and Bellotti, 1997; Simonsen and Kensing, 1997; Crabtree et al., 2000; Dourish, 2001; Randall et al., 2005) and shows no signs of abating.

No arbitrary boundaries were placed on the sort of fieldwork involved in this research. The fieldwork has involved interviews, both in workplaces and outside of them, observations of people at work and analysis of the products of work. The outcome of this research would not have been possible if the fieldwork methods had been restricted. Finally, the people and technologies studied have set the agenda for the analysis of the fieldwork. No pre-existing structure was imposed before beginning analysis; no pre-determined themes were looked for. This is in keeping with actor-network theory's directive to follow the subjects' ways of thinking about and organising the world and the relationships between the social

---

[1]By which Nanavati and Rajput seem to mean mobile telephones.

and "natural"[2] elements that make up their world (Callon, 1986b).

My primary concern has been the design of ASR systems for the workplace and it is that concern which has influenced my fieldwork and analysis. As Randall et al. (2005) say:

> What to look for, how to look for it, and how to assess its significance are, from the outset, design-related matters.

This may seem at odds with my assertion that there were no pre-determined themes in my research but it is not. There were no pre-determined themes for analysis but there was an over-arching orientation towards design.

This chapter takes the stance that there are a wide range of methodologies that have been developed for looking at and attempting to understand people, work and technology and since this thesis is concerned with design arising from fieldwork this chapter does not present an exhaustive account of all possible social research methodologies that could have been used. Because many of the techniques used in this thesis are practical in nature, they are best explained with examples of their use by others and then their use in this research.

Instead of creating a new method for conducting and analysing field research of technologies and work practices, this thesis uses existing analysis methodologies as a set of lenses. In looking at the case studies through each lens a different view is obtained and different properties are brought into the foreground for viewing. In later chapters the different views afforded by the techniques described here are used to explain features of how ASR is used in the workplace and to move towards a design for a future speech recognition situation.

The techniques described here are used in chapters 6 and 7 of this thesis and their use in inspiring a design of an ASR system is reflected upon in chapter 8.

---

[2]or technical.

3.1   Fieldwork for Design

Performance of fieldwork is not difficult; finding locations in which to perform fieldwork is.

The initial fieldwork at the ACT Magistrates Court was performed through interviews with the Chief Magistrate in meeting rooms and in his office. The interviews were recorded in handwritten notes which were supplemented by notes and examples written by the Chief Magistrate during the interviews. The Chief Magistrate's notes were largely for my benefit and took the form of examples of the sort of work that he might produce on the bench. It was during one of the interviews with the Chief Magistrate that he allowed us to enter a courtroom and take copies of the large rubber stamps (see figure 5.4) used to speed the recording of sentences and outcomes.

The outcome of our initial meetings with the Chief Magistrate was the production of a proof-of-concept prototype which was demonstrated at a Court of the Future workshop.

Later fieldwork at the Court took place in the "back room" of the Court and required more negotiation on my part to gain access to the secure area. Once in the secure area of the Court's "back room" I was able to observe and interview those workers who were involved in some way with the process of communicating outcomes. I found the interviewees by following the magistrates' bench sheets and folders as they moved through the back room.

Gaining access to the ASR users in the public service was more difficult. Finding ASR users who used the software daily was difficult because they (obviously) are just normal workers—they don't advertise their difference. It was suggested that the people with chronic fatigue syndrome were high-frequency users of ASR so I made several unsuccessful attempts to contact the Chronic Fatigue Syndrome association. Eventually, I was put in touch with a speech recognition[3] trainer who very generously offered to allow me to place an advertisement asking for subjects in the newsletter

---

[3]"Speech recognition" and "voice recognition" are terms used interchangeably by users of ASR dictation software.

she produced for clients. It was in this way that my interview subjects self-selected. I was able to make contact with six ASR users through the trainer's newsletter.

After being contacted by the ASR users through the advertisement I arranged to meet the interview subjects at a location chosen by them. I met one in her house, one in a cafe near her office and the others invited me in to their workplaces. All the interviews were recorded on audiotape and were augmented with handwritten notes made during the interview. Interviews typically lasted for between 45 minutes and an hour. There were only two planned questions that the interview subjects were asked:

1. How did you come to be using speech recognition? and,

2. What are your experiences of using speech recognition?

Other questions were asked only to prompt the interviewees to elaborate on answers that they gave in response to the planned questions.

In contrast with the effort involved in finding ASR users, making contact with the Hansard department was serendipitously simple. An article appeared in the Canberra Times on how the Hansard department at Parliament House used ASR. Contact was made with the manager of the section and meetings were set up. Once an initial meeting had been made I arranged time for observations and interviews with the Hansard editors. The interviews at Hansard were conducted at the first meeting in a meeting room in Parliament House. This was then followed by a tour of the offices and a demonstration of how they used the dictation software. A second interview took place in an employee's office. The interviews at Hansard were recorded on audio tape and augmented with handwritten notes. As well as asking work process-related questions, for example, "How do you use the speech recognition software in your work?", the Hansard employees were also asked about their experiences in using the software in general.

Interviews were transcribed initially by listening to the tape-recordings and typing, however after the initial meeting at Hansard I was inspired to emulate their "respeaking" technique and from then on I transcribed interviews by listening to the audio and simultaneously repeating what all parties to the interview said into Dragon NaturallySpeaking version 6 (Dragon NaturallySpeaking Preferred 6, 2002). Respeaking considerably improved my rate of transcribed words per minute.

Once the interviews were conducted and transcribed it was possible to analyse them using a variety of methods, as described in the next section.

3.2   Analysis for Design

Having conducted the interviews and observations, performing analysis of the data was the next step in my approach.

Analysis of fieldwork data for design in the area of computing systems has led to interesting results. For example, spreadsheets (Nardi and Miller, 1990) were studied and it was found that software that was assumed to be mainly used by a single user was actually used collaboratively. This was doubly interesting because there were no explicit collaborative mechanisms in the spreadsheet software and all the collaboration was managed by the users. Rouncefield et al. (1994) looked at the collaborative ways of working in a small office and the importance of the artefacts of work were shown to be equally important in the small office as they are in larger, more "sophisticated" work settings.

Ethnographic data, that is, field work data, is also increasingly used in commercial marketing research (Bruner, 2005) and product development (Blythin et al., 1997) to inspire and inform the design of many different products from Recreational Vehicles to breakfast cereal (Squires and Byrne, 2002). In the commercial world, even more so than in academia, incorporating ethnographic insight to the design process is problematic because of the contrasting and often incompatible goals of ethnographers, designers and business people (Blythin et al., 1997).

The canonical example[4] of ethnographic data being analysed for design in Computer Supported Cooperative Work (CSCW) is that of control rooms (Bentley et al., 1992; Heath and Luff, 1992; Hughes et al., 1992b,a,

---

[4]Canonical, at least, according to Dourish (2001). Air Traffic Control has been the subject of many ethnographic investigations.

1995) where the analysis showed that the use of various non-technical artefacts was essential to the work done in the control room.

More recently, the use of mobile phones by teenagers has attracted
much ethnographic interest (Taylor and Harper, 2002, 2003; Berg et al.,
2003; Satchell, 2003; March and Fleuriot, 2005) and revealed how teenagers
use mobile phones to establish and maintain a sense of identity, manage
relationships and maintain a sense of privacy in public spaces.

What the research described above has in common is that the leap from
raw data to insights from analysis is often murky or not described at all. In
computer systems design literature that uses field work as part of a design
process this lack of explanation could be because the analysis method for
raw qualitative data is seen as "messy", soft or merely conventional because of its long established tradition in other fields. Programmers don't
describe their variable naming scheme if it's in keeping with established
convention, for example. One of the most well known and well used methods for analysing qualitative data is grounded theory (Glasser and Strauss,
1967). To perform my grounded theory approach I read and re-read the
interviews and observations, comparing and contrasting them, allowing
recurring themes to emerge. By "working the data" in this way, the theory
for analysing the data emerges from the data inductively rather than de-
ductively (Fitzpatrick, 2003). The trajectory of using ASR in organisations
described in chapter 6 emerged from the interviews in this way.

Grounded theory aids in understanding data but does not need to
stand alone. In this research, two different approaches have been used to
further "work the data" in order to obtain deeper understanding and insight into the problems of using ASR in an organisation. The approaches
used are the locales framework (see section 3.2.1) and actor-network theory (see section 3.2.2).

3.2.1 Using the Locales Framework

The Locales Framework (Fitzpatrick and Kaplan, 1998; Fitzpatrick et al.,
1998; Fitzpatrick, 2002, 2003) is a package of methods derived from specific
sociological approaches, mainly Strauss's *Theory of Action* (Strauss, 1993),

presented in a way to make sociological insight accessible to Information Technology (IT) researchers. As such, the Locales Framework is focused on the *actions* performed by people in particular *locales*.

In the Locales Framework (Locales), a locale is the primary unit of analysis. A locale "is the place constituted in the ongoing *relationship* between people in a particular social world and the 'site and means' they use to meet their interactional needs" (Fitzpatrick, 2003, p.g. 90). Site and means are the physical space as well as the resources available there, resources being tools, objects, and so on. There are two phases to Locales, relating to either understanding or designing, though in practice both are interlinked. In the understanding phase Locales directs the researcher to look at how the people in the locale use the resources at their disposal. The understanding phase emphasises the situated, complex, dynamic nature of work. The designing phase of Locales is described in the next section, 3.3.

There are five aspects to the Locales Framework: locale foundations, civic structure, individual view, interactional trajectory and mutuality. Locale foundations identifies the social world of interest and civic structure takes the social world of the locale into the wider sphere of the social, political, organisational, and so on. Individual view deals with the different people involved in a locale and their views on the locale from each of their positions and roles. Interaction trajectory is about actions and ways to achieve actions. Finally, mutuality is concerned with the shared achievement of awareness in the locale. The analysis in this thesis relies primarily on the aspect *interaction trajectory*.

> The *interaction trajectory* aspect captures all of the dynamic and temporal aspects of the living social world and its interactions within and across locales—past, present and future—and the co-evolution of action, locale and social world as the trajectory unfolds. (Fitzpatrick, 2003, p.g. 120)

The interaction trajectory aspect of Locales is particularly concerned with the *situatedness* of work and action, a concern that has been a theme of other work in the broad field of CSCW, (see for example Suchman, 1987;

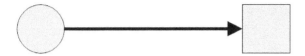

Figure 3.1: A simple trajectory with a single start and end point.

Robertson, 2002), and the importance of work and work practice. Fitz-patrick says that interaction trajectory is about how people interact with the setting and each other and how they evolve a new set of conditions for further action. This means that interaction trajectory is concerned with past, present and future action as well as the processes by which actions evolve. Further, it looks at the people involved in the interaction, the "in-teractants", who all bring their own histories and biases to the interaction, a concept that is similar to *Technological Frames* described by Orlikowski and Gash (1994).

Interaction trajectory "captures the dynamic temporal aspects of the so-cial world interactions in and across locales" (Fitzpatrick, 2003, p.g. 130). Interaction trajectory is also applicable at many different levels of detail. In analysing the use of ASR in the workplace, interaction trajectory was useful in understanding the past, present and future of each users' inter-action with their ASR software. In this thesis interaction trajectory can apply at an individual level to explain the trajectory that a single user has followed in their use of ASR software (see figure 3.1). And it can also ap-ply at a more nuanced level where broadly similar starting locations lead to broadly similar finishing locations in each users' experience of the soft-ware (see figure 3.2, p.g. 46) which is not to deny the uniqueness of each users' experience but to distill specific experiences into generalised, use-ful, properties of the "site and means" that can be applied elsewhere.

The concepts of past, present and future play a part in my analysis of the work of the ACT Magistrates Court though as that work is much more structured the concepts of phases, rhythms and schedules are more useful. These concepts can also be used to focus on coincident trajectories and the relationships between trajectories which I have used in looking at the distributed work of the many workers at the Court.

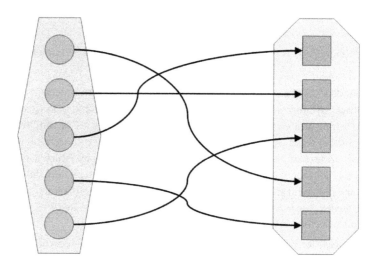

Figure 3.2: Many trajectories, originating and ending in different locations, though having much in common.

Interaction trajectory addresses many specific aspects of a *locale*. Using interaction trajectory for analysis first requires a focus on a particular interactant. The focus of the analysis is the entity or entities which are the subject of the trajectory. Entities can be single or multiple, simple or complex, tangible or intangible, obvious or obscure. An entity may be human or non-human, a social world, an object, an event or a concept or idea. The interactants are the agents undertaking actions or interactions that make up the trajectory. The primary question asked of interactants when using Locales to understand a trajectory is "what role does the interactant play in relation to the focus and how do they carry out that role?" (Fitzpatrick,

2003, p.g. 122) In performing an analysis of my data, I have asked this question of ASR software, users of such software and the other interactants such as files and stamps I have identified in the locale in which this use takes place.

Being concerned with interactants, Locales is also concerned with the actions and meta-actions they undertake. Trajectories as courses of action are made up of different sorts of actions. Work actions and meta-actions relate respectively to work, and work done to allow work to be performed. For example, Robertson (1997) describes the work and articulation work necessary for software designers to work in groups.

Fitzpatrick (2003) identifies conditions for action and support for action as being related to interaction trajectory though such conditions and support also overlap with the locale foundations, civic structures and individual view aspects of Locales. In my analysis of ASR users and the Court, I treat conditions and support for actions as being related to interaction trajectory and specifically the site and means as being conditions and support for various actions undertaken by the various actors in those locales. At the Court it was particularly useful to consider the role of artefacts in the coordination of work, a stance that is also reflected in the (pre-Locales) analysis of work done in control rooms for air-traffic control (Bentley et al., 1992) and in the London Underground (Heath and Luff, 1992).

Of particular interest to me, and of particular use in my analysis of the use of ASR in the workplace, was the "co-evolution of action, locale and social world as the trajectory unfolds" (Fitzpatrick, 2003, p.g. 120) which is not really specifically addressed in Locales but which informs, in particular, the notion of interaction trajectory. Actor-network theory addresses the co-evolution of interactants in a locale, or in actor-network terms, actors in a network in a very interesting way and is the subject of the next section.

3.2.2 Using Actor-Network Theory

Actor-network theory (ANT) is a theoretical framework that comes from a field called "social studies of science and technology" (or, science and

technology studies (STS)). It was not intended to be used to describe information technology, though it has been used in that way, and was certainly
not intended to inform the design of a computer system. ANT is, however,
flexible enough that its outlook and vocabulary can be translated for the
purposes of analysis for design.

ANT has been used for analysing the interplay between many different social systems and technologies from scallops and fishermen (Callon,
1986b) to electric vehicles (Callon, 1986a), the Portuguese expeditions to
India for the spice trade (Law, 1986, 1987), seatbelts (Latour, 1992) and alternative public transport systems (Latour, 1996). ANT has more recently
come to the attention of computer systems researchers (Hanseth and Monteiro, 1998; Tatnall and Gilding, 1999). What all of these disparate analyses
have in common is the language of ANT, which is particularly geared towards describing "the small, concrete technical and non-technical mechanisms which go into the building and use" (Hanseth and Monteiro, 1998)
of socio-technical systems. ANT is useful for explaining how technologies, and groups of technologies and people, are incorporated into new
situations.

The *actor* in ANT represents many things. In an example of driving a
car, driving is influenced by the type of car, whether it has automatic or
manual transmission, the driver's previous experience in driving, the pre-
vailing traffic conditions and so on (Hanseth and Monteiro, 1998). The act
of driving a car takes place within and is influenced by all of these factors. ANT says that all of the factors should be considered together. An
actor-network is the act and all of the factors influencing the act linked
together in a network, which are themselves linked to other acts and factors. Since in these terms everything is eventually linked to everything
else, performing an actor-network analysis is partly about deciding which
act and which factors are going to be analysed. It is difficult to say before undertaking an actor-network analysis where the analysis will lead
as influential factors emerge during analysis.

The ANT research frame (Callon, 1986b) directs the researcher's gaze
towards particular aspects of a socio-technical system that are useful dur-

1. Interessment

2. Enrollment

3. Points of Passage

4. Trial of Strength

Figure 3.3: The Actor-Network Theory Research Frame developed from Callon (1986b).

ing analysis. It comprises four inter-related overlapping steps that describe how stable actor-networks come to be established (figure 3.3). Stable actor-networks are of research interest because they represent the *status quo*. The research frame can be used to ask how stable networks became stabilised. The failure of an actor-network to become stabilised can equally be examined by the research frame.

In the *interessment* step, actors are made interested in joining an actor-network. The way in which individual actors are interested is unique to the particular actor-network. In the *enrollment* step, actors agree to play a role in the network, they are *translated* into the network and are *inscribed* with a *program of action*. Put another way, actors who join a network are given a script to follow. In any network, one or more actors attempt to establish themselves as a *point of passage*. The point of passage is the actor who assigns roles to, or literally represents[5], the other actors in the network. Conflicts arise in an actor-network when more than one actor attempts to establish themselves as a point of passage. In the *trial of strength* it is seen whether the actors adopt the roles assigned to them.

The steps in the research frame are not linear but are, in every actor-network, constantly in play. The relative stability of an actor-network determines the work required to establish a new actor in the network. Stable networks are almost self-sustaining, requiring little effort on the part of the actors who have a stake in the network's continued existence. Unstable networks require much effort to sustain in the face of other competing networks attempting to dis-enrol actors from the less stable network in

---

[5]Represents or acts as spokesperson for.

order to have the actors not follow their assigned scripts, making the network fail its trial of strength or leading to the break up of the network.

For the purposes of analysing the qualitative data in this research, the most important concepts from actor-network theory are *inscription* and *translation*. Inscription and translation are intimately tied to design. For the purposes of analysis in this research, these concepts are used to explain how a designer inscribes a program of action into a software product—dictation software.

Akrich (1992) describes inscription as the act of a designer:

> Designers thus define actors with specific tastes, competencies, motives, aspirations, political prejudices, and the rest, and they assume that morality, technology, science, and economy will evolve in particular ways. A large part of the work of innovators is that of "inscribing" this vision of (or prediction about) the world in the technical content of the new object.

However, the inscription of a program of action into a software product does not guarantee that the users will follow the program. Abdelnour Nocera and Hall (2004) describe the conflicts that arise when users expect a different program of action to the one inscribed in ERP[6] software—both the users and designers think that their program of action is correct. To ensure that users follow the designer's inscribed program of action, they must be translated, i.e. their interests must be reinterpreted and re-presented in order to be aligned with the inscribed program of action.

The processes of inscription and translation do not end when a software product is shipped. Building and maintaining an actor-network is an ongoing process. Inscription and translation are performed by users in the process of using the software and making it useful. Because each user's work is slightly different, according to actor-network theory, they must translate the dictation software to re-align it with their work process. In doing so they inscribe a new program of action, i.e., their work process, on the software.

---

[6]Enterprise Resource Planning.

The strength of the actor-network inscribed into the software by the designer influences the ease with which a user can change the program of action and so enroll the software in a new actor-network. The degree of difference between the program of action inscribed into the software and the program of action that the user wishes to inscribe will also influence how easily users can translate the software to their ends.

The analysis that ANT allows of the use of ASR software is to show how users translate the software and how they strengthen their inscription in order to make the software useful to them.

The ease with which the new actor-network is established and maintained is related to the flexibility of the existing actor-network in which the users are working and the flexibility of the software.

Because there are two case studies in this research, the injured users and the Hansard users, and both use what is ostensibly the same software, identifying how the software is translated, how the users' work is translated and how other actors are translated in order to make the software usable can be done by comparing and contrasting the different cases.

3.3 The Emergence of Design from Analysis

Analysis of fieldwork is about explaining the present. Design is about inventing the future. There is a fundamental dichotomy, therefore, between accounts and analysis of the present situations and the process of designing future situations. Making the leap between the present and the imagined future is problematic. Hughes et al. (1992b) said that the CSCW[7] community was "faltering from ethnography to design" and proposed ways to incorporate social understanding with technical design (Hughes et al., 1992a, 1995). More recently, Button and Dourish (1996) and Dourish and Button (1998) proposed *technomethodology*, a fusion of ethnomethodology and design, as a way to move from critiquing design to doing design. Crabtree (2004) extended technomethodology and provided a set of

---

[7]Computer Supported Cooperative Work. CSCW, more so than other areas of computer science, has made the most use of field work methods to understand how people work with computers.

steps with which to frame the ethnomethodology/design process. Tech-nomethodology was not applied in the process of completing this research however it is considered retrospectively in chapter 8.

Many of these approaches have been developed because despite the desire of designers to use fieldwork (or ethnographic) data, ethnographers do not typically communicate their findings in ways that designers find useful. Some have argued that moving from ethnographic methods to design is fraught with difficulty and danger (Suchman, 2003) and that care must be taken lest ethnographic (anthropological) insights are used against people instead of assisting them. In this thesis, at least, the danger has been averted as much as possible by having no outside influence other than the research impulse.

Randall et al. (2005) have said that analysis and design are inseparable and that has certainly been the case in this research. The split between analysis and design described here is primarily for clarity of explanation.

The process of design, even more so than analysis, is complex. Even more so than my performance of analysis, my performance of design has adopted methods and approaches as it became necessary to make the leap from a particular analysis to a design product. This section describes the techniques I have used to arrive at the design described in section 7.3.

In explaining how I went about designing for the ACT Magistrates Court, I have further split the process into concepts I have "built in" to the design and methods used to arrive at a (more or less) concrete design.

3.3.1   Concepts Used in Design

This section is necessarily brief as it is difficult to describe the concepts used without describing their use. The broad concepts used to influence design in this thesis have come from the Locales Framework and from Actor-Network Theory.

Concepts from the Locales Framework

Using the Locales Framework to design is primarily an exercise in explor-ing ways to enhance locales by using existing spaces and resources differently or by evolving entirely new locales. The use of Locales for design is "driven by interactional needs and understanding the broader context(s) in which those interactions happen" (Fitzpatrick, 2003, p.g. 149). In order to use Locales in this way, questions are asked during the design process that stem from Locales itself. Broad questions addressing the interactional needs of the locale are asked, such as: how could the interactional needs be supported? How could technology enable new ways to interact?

Designing is about the exploration of possibilities. Using Locales to design is "about exploiting the strengths of any available medium [...] to better meet the social world needs" (Fitzpatrick, 2003, p.g. 149) identified during analysis while simultaneously being aware of the essential properties of the existing locale and the interaction(s) that take place there.

The influence of Locales on the design work in this research has been to direct the imagining of technology in ways that improve or enhance work. This has entailed identifying which components of work at the ACT Magistrates Court are able to be changed and which are important not to change. Having determined where changes could be made it was possible to design new ways of working inspired by the *interaction trajectory* aspect of Locales. In Locales' terms this has meant finding whether actions are work-actions or meta-actions; whether they are independent of others or interdependent; whether they are routine or not routine; and so on. Also identified using Locales were the "site and means" of the locale. Questions were asked if it was possible to change the site and means and how the resource would enhance the locale or the work done there. Finally, using Locales in the analysis phase allowed the identification of important aspects of the work done in the Court and how it was done such that those aspects were, as much as possible, left intact during design, having been identified as an important part of the work and meta-work at the Court.

Concepts from Actor-Network Theory

Using actor-network theory to design, i.e. to imagine the future, is using ANT in a way that it was never intended to be used. However, this does not make it impossible. ANT is flexible enough in use to be translated from an analysis tool to a design tool.

Using ANT to analyse a situation means identifying the actors and how they are enrolled in a network. Describing the power structures, the points of passage and the flexible and inflexible actors allows a story to be told of how a particular actor-network came into existence and how that existence is sustained. Designing with ANT follows a similar path.

As Akrich (quoted in section 3.2.2) said, "A large part of the work of innovators is that of 'inscribing' this vision of (or prediction about) the world in the technical content of the new object". The new object in this case is an ASR system for the Magistrates Court. The work of design in this case, then, is to inscribe the vision of the ACT Magistrates Court using ASR into a system.

The ANT Research Frame (see figure 3.3 on p.g. 49) can be used to guide design. Having broadly imagined the future, the designer can then begin identifying the actors who are essential to the success of the new system. By interesting these actors in the new system and then enrolling them, that is translating their interests to bolster support for the new system, the designer can begin to strengthen the vision of the future and be more sure that it will come true. Because the designer is trying to establish the new system as an obligatory passage point they must identify other actors who would oppose the new system and find ways to translate and enroll the opposing actors in the future actor-network. Finally, the designer must imagine the system in use and show how the system and the actor-network of which it is part, holds together with all of the actors playing their assigned roles.

Obviously, the more novel the system is and the more divergent the imagined actor-network is from the existing actor-network the harder it is for the designer to translate the interests of the actors so that they will follow the new program of action that the designer will inscribe on the new system. In performing design influenced by ANT I have attempted to follow a path of least resistance by imagining roles for the actors in the new system that as closely related as possible to their existing roles. By keeping the new imagined roles as close as possible to the existing roles it can be assumed that the participants will be more likely to follow them.

This is in keeping with the actor-network approach.

3.3.2   Ways to Design

Having introduced the concepts used in thinking about new designs for ASR at the ACT Magistrates Court in the previous section, this section introduces the actual concrete actions that were performed to generate the designs.

Crabtree's 2004 Technomethodology, while not explicitly used in conducting this research, can be considered a pattern for conducting ethnographically-inspired design research. Technomethodology, as conceived by Crabtree, is an spiral-model process, beginning with the construction of a novel technology which is then released into the "real world". The effects of that release are observed by the field workers who then "explicate the accountable structures of practical action made visible" (Crabtree, 2004) in the release. The findings of the field work are then fed back into the construction of a new or refined technology. In this thesis, the use of ASR in the Public Service is treated as the release of technology into the world and the research that takes place is the considering of how that release has affected the Public Service users and the "practical actions" that they undertake in order to use the technology successfully. These findings, along with field work conducted at the Magistrates Court, are fed into a design for ASR at the Court. The ways that the technomethodology model relates, and does not relate, to the research conducted in this thesis are examined in chapter 8.

Many of the approaches used in this thesis to generate new designs have their roots in the work of Campbell et al. (2003) particularly the design tools of the "circus" and the various representations that Campbell et al. describe. A design circus is a way of exploring representations of the situation being designed for and artefacts from the situation as well as presenting ideas for designs that arise from the situation and its artefacts.

In the design circus that was held to aid in inspiring design for this work, a representation of the ACT Magistrate Court's work flow was produced so that participants could understand the process that was being considered in the potential new design. Anonymised bench sheets and imprints from the rubber stamps that the magistrates used on the bench were presented, to illustrate the documentation that was produced on the bench.

The participants in the design circus were representatives from various groups who were interested in ASR at the Court. Lawyers, computer-systems designers and interaction designers were present. The Court was invited to send representatives but did not respond to the invitation.

The design circus contributed to the prototyping of a new design for ASR at the Court.

ASR presents unique challenges when producing prototype designs, particularly early prototypes that are intended to stimulate design think-ing rather than be seen as design proposals. A graphical user interface (GUI) can be prototyped using various low-fidelity methods, even to the extent of prototyping in paper (Snyder, 2003), however in ASR interfaces, much of the interaction, invisible and impermanent as it is, is difficult to represent on paper using methods derived from GUI design. Other as-pects of the speech user interface, for example vocabulary and grammar, are too detailed to specify at the prototyping stage.

The method used in this thesis to prototype a speech user interface is to use detailed scenarios (Carroll, 1995) to describe the ASR system in use. Technical constraints are incorporated into the scenarios by basing them on pre-existing technical research. Given the methodological stance of this thesis regarding analysis of qualitative data, this approach to design allows the richness of the imagined future to be described in a similar manner to the lived present.

The scenarios used for design are inspired by those described by Bødker (2000) where scenarios are paired in positive and negative caricatures of action. Instead of a single scenario that describes imagined future ac-tion, Bødker's scenarios allow description of action as well as showing the imagined positive and negative effects of the design upon the situation.

3.4 Summary

The methods described above are used in this thesis as a way to approach analysis of fieldwork for design. These approaches are introduced in this chapter so that the reader is not surprised when various terms and concepts are used in later chapters. This chapter also serves to locate this thesis in the wider sphere of sociological analysis of the use of technology.

The next chapter (chapter 4) is an ethnographic account of the workplaces where I encountered ASR in use. The subsequent chapter (chapter 5) is an ethnographic account of the ACT Magistrates Court, a workplace where ASR could be used in the future. Chapters 6 and 7 apply the concepts introduced in this chapter to the ethnographic accounts.

CHAPTER 4

Automatic Speech Recognition Users

The empirical basis of this thesis is three case studies. Two of the case studies are of real work situations where commercial off the shelf automatic speech recognition (ASR) is used productively and the third is that of a workplace where automatic speech recognition could be introduced, the ACT Magistrates Court (see chapter 5). The two real work situations are the Hansard department at Parliament House (see section 4.1) and an amalgam of six different users of ASR who work for the Australian Public Service (APS) (see section 4.2). This chapter presents rich descriptions of the two situations where ASR is used. The next chapter presents the situation where ASR could be introduced.

I gained access to the Hansard department through an initial contact made by one of my supervisors. The APS automatic speech recognition users who generously volunteered their time were found through a speech recognition trainer.

This chapter describes each case study in a scenario-like manner using a persona to represent the core character of each scenario. Describing each situation in a scenario allows the situation to stand alone, allows the interviewees to remain anonymous and, in the case of the APS speech recognition users, allows me to generalise the interviewees' fairly diverse work

practices into a single representative story. As all of the interviewees perform office work, their work practices are diverse in detail, though all deal
with documents and communicate with colleagues. The APS scenario is
constructed from elements of each of the interviewees stories allowing the
scenario to be representative while preserving the subjects' anonymity.

4.1   The Hansard Department

At Australian Parliament House, the Hansard department (properly called
the Department of Parliamentary Reporting Staff) is concerned with producing the document called Hansard that is a record of what was spoken
in the House of Representatives, the Senate[1] and in various committees.
The current work practice has some Hansard editors using commercial
ASR software to re-speak the words of the parliamentarians to transform
their speech into text that can be edited and formatted into the document
called Hansard.

   The editors have a great deal of freedom in how they use the ASR software. They may only use it for re-speaking and do all of their editing with
mouse and keyboard, or they may do some editing by hand and some by
voice, or they may do all of their editing by voice. Hansard, the document,
is termed by the Hansard department a "rational verbatim" document
meaning that it is not a perfectly accurate representation of the parliamentarians' speech but that hesitations, false starts, non-speech sounds and
other noises are not included in the document that is produced. Within
the Hansard department, ASR software is treated as another business tool,
in much the same way as laser printers, monitors or spreadsheet software
are treated in other businesses or government agencies.

   The scenarios that follow are entirely fictional. There may be a Hansard
editor named Paul but the scenario does not describe him, his opinions or
his work practices. The scenario is an amalgam of my observations of
the work process in the Hansard department at Parliament House and is
meant to illustrate the work environment that I observed. It may contain

---

[1]Australia's parliament is modelled on the Westminster system with an upper house,
the Senate, and a lower house, the House of Representatives.

inaccuracies and even mistakes about the process — I take full responsibility for any misrepresentation of the Hansard department.

4.1.1 Scenario: Paul the Hansard Editor

Paul is a Hansard editor. He works in Parliament House in Canberra in the same building that contains the House of Representatives and the Senate. The Hansard department's offices are on the ground floor of Parliament House which is quite prestigious — other sections have their offices in the basement or even in other buildings in the Parliamentary precinct. Paul quite likes working in Parliament House and is glad that the office he shares with Janelle, Phoebe and Melanie has a window.

Paul's job is to create the document called Hansard. People think that Hansard is just what's said in the House of Representatives and the Senate but it can also be a record of the various committees that sit in Parliament House. Whatever the content of Hansard, the document itself has a short turnaround time and has to be completed and edited for the next day's session. Sometimes one of the Houses or committees can run late into the night so Paul and some of the other editors can leave work quite late in the evenings.

PAUL'S OFFICE. The office that Paul shares (see figure 4.1, 62) with his colleagues is a fairly large square room with four desks in the corners and shelves over the desks. Standard equipment is a powerful computer with a large monitor and what seems to be a tangle of black headphone cables leading to a black box next to the computer. The editors also have a standard issue Australian English dictionary and several other reference books and memos on Australian place names and other topics that might come up in Parliament. Paul, like all of the editors, has two sets of headphones, a high quality set for listening to audio and a high quality headset microphone that he uses for speech recognition.

When Paul comes in to work it's usually about 8.15am. He sits down at his desk and calls up the schedule of what he has to edit today. Like every other editor, Paul is assigned seven minute long parts of the proceedings of the House or the Senate to prepare for inclusion in the larger document

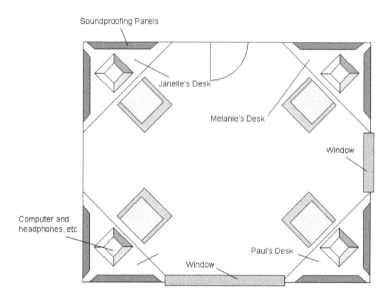

Figure 4.1: Plan diagram of Paul's Office.

that is the day's proceedings. Paul's first session starts at 8.37am in the House of Representatives. It's only 8.21am so Paul has time to check his email before having to leave the office to walk to the House of Representatives to observe his first scheduled session. Leaving the office Paul sees some colleagues as they arrive. He exchanges quick "hellos" with them but they're used to editors leaving to observe their assigned sessions and they don't expect him to chat.

THE CORRIDORS.   Walking down the corridors to the Reps (as the editors call it) Paul sees more people that he knows. Parliament House is a large building and there are people dressed in gym clothes who are arriving at their offices from their morning workout. Paul also sees one or two members of the Reps dressed for exercise. As he approaches a large intersection of corridors Paul smells coffee from one of the cafes in the building. He'd love a coffee but it's always better to be early for a session than just on time. When Paul first started in the Hansard department the size of Parliament House was a little overwhelming but he's used to it now. If you're not familiar with the building it's easy to get lost but there are some very subtle cues built in so that you slowly become able to orient yourself. Things like different coloured carpets in different areas help and also the prominent artworks that serve as landmarks. It's different to working in a traditional office building as Parliament House is only three stories high, if you count the basement, but it's very spread out. Walking from the House to the Hansard offices and back again several times a day is good exercise.

INSIDE THE HOUSE OF REPRESENTATIVES.   At 8.33am, Paul sits next to Janelle who is just finishing the first session for the day. Janelle is sitting in front of a laptop computer, typing notes in a file to aid her recollection of the last seven minutes when she gets back to her office. Because the Speaker of the House controls the Members' microphones, only one person is allowed to speak at a time, and only the active microphone is recorded. When people interject or try to talk over the Member who has the floor their interjections aren't picked up clearly on the audio so the editors make notes to remind them who interjected. When they come to create the official record, most of the interjections are left out, unless they contribute to the clarity of the record.

Janelle stands up and leaves and Paul sits in front of the laptop. They exchange smiles and say hello but the House doesn't stop for chit-chat and they'll get to talk more over lunch. The Member for Bennelong is introducing a new Bill to Parliament which means that he's reading quite a long speech. There's not much else going on this early in the morning, just the occasional "hear, hear" from the government side, so Paul doesn't have to make too many notes. At 8.43am Phoebe comes up and they swap places at 8.44am.

BACK IN THE CORRIDORS.   Walking back along the corridors (greeny-blue carpet, past the abstract statue that looks like a nose), Paul sees his friend Jared who works in the basement in the broadcast section. Jared says, "I'll email you about the footy-tipping competition," as they pass each other. They're too busy today to stop for a chat.

Paul is past the cafe when he remembers he wanted a coffee. He can't bring himself to backtrack so his morning pick-me-up will have to wait.

PAUL'S OFFICE.   Back in the office, Paul selects the session he just came from in a custom piece of software that displays all of the time slices of the House and the Senate and whatever committees are sitting that day. A small window pops up that looks like the MP3 software that Paul uses to listen to music on his home computer. Before getting down to actually transcribing the session, Paul puts on his headphones and listens to the session once more so that he can anticipate some hesitations or mispronunciations that the minister made. To begin transcribing, Paul swaps from the headphones to the headset microphone and starts the speech recognition software using an icon on his desktop. Paul listens to the speech through the headphones and repeats every word, leaving out any hesitations or mispronunciations as he hears them. They're not supposed to appear in the final document and it saves time to do a bit of editing on the fly.

When Paul started with the Hansard department he was a little wary of the speech recognition software as it was new to him and didn't seem to respond well to his voice but with some practice and help from the more experienced editors Paul is now proficient and rarely has problems. The hardest part of using the speech recognition software is remembering to speak the punctuation and deciding where to put it. One of the more usual instances where Paul has to speak punctuation is to bracket all of the times the Member of Parliament addresses the speaker. The MPs have a habit of saying something like, "Mr Speaker, it is quite clear that the Member for Oxley, Mr Speaker, does not really understand the current situation". Paul thinks they say "Mr Speaker" instead of saying "umm" or "ahhh". Every time someone in Parliament uses the Speaker as an excuse to pause, Paul has to say: "Mr Speaker *comma* it is quite clear that the Member for Oxley *comma* Mr Speaker *comma* does not really understand the current situation". Paul will also attempt to make some grammatical sense of the speech he is re-speaking by constructing it into proper sentences. This is usually easier when the MP is speaking from notes or a prepared speech but re-speaking question time when they are speaking off the cuff really shows you that people rarely speak in correct sentences!

The software that Hansard uses is based on *Dragon NaturallySpeaking* (Scansoft, 2005) but it's been heavily customised for Hansard's purposes. There are lots of shortcuts that the editors have developed, in conjunction with their IT team, to make transcribing faster. The completed Hansard document looks much like a television script or a play and as Paul respeaks the speech he uses the custom short-cuts to create the basic script-like structure. He's tried creating a Hansard document on his home version of the software and creating the formatting is almost impossible to do by voice so the short-cuts are a real boon to productivity. The short-cuts are called 'macros' which is an IT person's way of saying they're a collection of individual commands that you can complete by saying one command. The most useful macro is the one that does the special formatting for Hansard in one command. The special formatting does a lot of things but mainly makes the text look a lot like a script with very little effort on Paul's part. When Paul says "Go speech Bennelong" the macro puts:

MR HOWARD:

on the screen. Bennelong is Mr Howard's seat, and this command is actually easier for Paul because the Speaker of the House addresses Members by their seat, not their name.

Paul can then repeat Mr Howard's words. When another Member starts speaking, Paul can say: "Go speech Eden-Monaro" and what the Member for Eden-Monaro says and the screen shows the previous transcription and the result of the new command:

MR HOWARD: Mr Speaker, I call on the Member for Eden-Monaro to report on the budget.
MR NAIRN: Thank you Prime Minister.

The macro makes it possible for Paul to do the SMALL CAPITALS for the Member's name as well as other more subtle formatting without explicitly speaking the commands for small capitals.

Paul finishes his re-speaking transcription of his seven minute slice. While he was re-speaking, Phoebe returned from her first session of the day and Janelle finished re-speaking her session. Having three or four people in the one office re-speaking isn't a problem as the headset microphones are quite sophisticated and can cancel out a lot of the noise that isn't coming from very close to the microphone arm. The office is also laid out to minimise noise, including having new sound deadening panels in the corners. Everyone learns to work quietly after a few weeks, too.

Paul swaps headphones again, back to his audio-only pair, to check through what he's just transcribed. The headset-microphone pair are not as comfortable and the sound quality isn't as good as the audio-only set. Doing a final listen to compare to the transcription shown on the screen with the audio record is valuable for picking up any errors that the speech recognition process has made, as well as checking that he didn't miss anything. Paul has a foot pedal for controlling the audio which lets him type as he listens and manipulates the playback. When he first started with Hansard he made lots of mistakes doing transcription but he's become much better lately and is noticing that he's not correcting so many re-speaking mistakes. Speech recognition errors still occur, especially later in the day when he's a little tired and he gets a little sloppy with his pronunciation — not that anyone listening would notice, but the software seems to be quite sensitive to such small variations. When he's got a cold it sometimes seems like it's not worth the effort to use the speech recognition system to do the transcription, it makes that many errors. He's thought about creating a special voice file for when he's got a cold, but the training procedure is pretty slow and he doesn't get sick often enough to get enough training done to get to a productive level.

The kind of errors that the speech recognition software makes are very strange and Paul is often surprised by the strange things the software does. Sometimes, when he says "go speech Eden-Monaro" the computer doesn't run the macro but simply writes "go speech eden monaro"[2] on the screen. According to the IT guys, this happens because Paul said the command

---

[2]The software doesn't even capitalise the "Eden-Monaro" or put the hyphen in. Often it writes something else entirely.

with *just* the wrong timing for it to be interpreted as a macro and it was interpreted as the three words instead. The other sort of error that the software makes is to get words totally wrong. Paul's favourite example of speech recognition software getting words wrong is to say "Australian" in just the right way to make the software write "astray alien".

After the final listen, Paul goes over the transcript again in a final check. It's surprising what you find, even after looking at the same words for so long. Recognition errors are different from typing errors and it took Paul, and all the editors, a while to be able to notice them. He's satisfied with the transcript, so he saves it and, using more custom software, uploads it to a central space so the compilation of the whole day's document can be started.

Paul checks his watch and sees that there's ten minutes until he has to go to the House of Reps again, just enough time to email Jared about the next footy-tipping round but still not enough time for coffee.

4.2 Injured Automatic Speech Recognition Users in the Public Service

Within the APS there are many office workers who begin to have pain in their hands and arms when typing or using a mouse. Popularly these injuries are known as RSI (Repetitive Strain Injury). A collective term for injuries of this type is "soft tissue injuries" as the specific injury may not be RSI but some other form of overuse injury or something altogether different from RSI. Many of those who acquire soft tissue injuries are lower-level office workers whose work involves high volume computer input and output though some of the people I interviewed were ranked quite highly.

When a worker first begins to feel pain in their arms and hands they may attempt to work around it, or adopt different methods to use their computer such as changing their seating position or the position of their keyboard and mouse on their desk. If ergonomic intervention, for example, changing a worker's seating position or desk layout, does not solve the problem they may take time off work for physiotherapy and their doctor or physiotherapist may recommend that they use ASR software to prevent the injury from re-occurring. Depending on the injured person's position in their organisation, this may be very easy to achieve or it may be difficult. Often, negotiating an injured person's return to work with ASR software involves the interaction of the IT department, the Occupational Health and Safety department, the injured person, their doctor or doctors, their manager, their rehabilitation officer, and possibly even more parties.

As with the Hansard scenario in the previous section, this scenario is made up and is an amalgam of many people's experiences.

4.2.1   Scenario: Kelly, the middle level manager

Kelly is a middle level manager in a large government department in Canberra. She is directly in charge of a small team of people and indirectly in charge of quite a large group. She reports to a section head, who reports to the department head. Kelly's work involves a lot of correspondence with other departments and sections, usually by email but sometimes by official memo. The group that Kelly heads mainly produces reports on various topics of interest to the department. The reports can be quite long, over thirty pages, and are quite time consuming to produce.

When Kelly comes into work, the first thing she does is put on a headset microphone and tell her computer to "wake up". Kelly uses speech recognition software to control most of what her computer does while she's at work.

A few years ago, Kelly started getting pain in her arms after long periods of typing. The pain steadily got worse and she had to take time off in the end for lots of physiotherapy. The physio eased the pain but did not remove it completely. After seeing a number of specialists, one decided that Kelly had what he called a "soft tissue injury" which has a technical name but Kelly explains it to her co-workers as "sort of like tendonitis and sort of like RSI". Kelly's soft tissue injury makes it painful for her to type or use a mouse for more than about 5 minutes. In some ways, her doctor tells her, she is lucky, because he has some patients who find it painful to write with a pencil or hold a knife and fork. Kelly's doctor recommended that she return to work and that she use speech recognition software to allow her to use the computer. Her doctor didn't really know a lot about speech recognition but Kelly's injury was bad enough that there was no other way that she could really return to work.

Kelly was very happy to return to work and because she considered herself to be fairly computer literate was happy to use the speech recognition software. Her manager was pleased that Kelly was going to return to work and arranged to have the IT department purchase and install the speech recognition software for Kelly. The IT section was reluctant to install the software for Kelly as it was not part of their carefully constructed "unified desktop", however Kelly's manager and the department's Occupational Health and Safety officer convinced them that it was for the best.

Kelly was lucky to work in a department that had an internal IT section. While they were reluctant to install the speech recognition software, they were able to be convinced to install it. After she had used the software for a while, Kelly heard of someone in her situation who worked in a department that had all of their IT support done by external contractors and that person had found it almost impossible to get the speech recognition software installed. Apparently when anything goes wrong with that person's computer the contractors always blame the speech recognition software.

The first thing that Kelly had to do after the speech recognition software was installed was to train the software to understand her voice. The training involved reading some set passages to the computer. After Kelly had completed training the speech recognition software to her voice she began using it to do her work. She was often frustrated as the software seemed to have a mind of its own, writing words that she didn't say and misrecognising words that she did. Her computer also crashed a lot more than Kelly was used to. She went to the IT section and asked them to stop the crashing because she suspected that the speech recognition software was too taxing for her current setup. The IT section said that they couldn't do anything for her as she was running software that was outside the "unified desktop" and that her computer met the recommended specifications for all of the software installed on it, including the speech recognition software. In the end, nothing was done.

Kelly isn't senior enough to have her own office and was even more junior when she first started using the speech recognition software. Her boss's boss doesn't even have an office, though he does have fairly tall partitions around two sides of his desk. Kelly likes the feeling of being part of a team and enjoys hearing the banter that goes on in the office throughout the day, though if the conversation gets too loud, it can sometimes interfere with the accuracy of the speech recognition software. Sometimes, if Kelly is reading over what she has "written" and someone else is speaking loudly, the microphone will pick up some of the sounds of their speech but in a strange way so that only nonsense gets written on the screen, usually something that looks like "the chat go of off on" or something equally nonsensical. Other sounds in the office that can affect the speech recognition software are ringing telephones and, if the office is very quiet, the noise from the photocopier can make a difference (though no-one believes Kelly about the photocopier).

Today, Kelly has an appointment with Julie, who is a speech recognition trainer. Kelly contacted Julie after hearing about speech recognition software trainers through an on-line discussion forum dedicated to people who use speech recognition in their daily work. This is Kelly's third appointment in six months with Julie. Julie has suggested that it might be the last time that Kelly needs to see her, as she seems to be having fewer and fewer problems.

For the first appointment, Julie came to Kelly's desk and watched her work and then made a lot of suggestions about different ways to work with the speech recognition software. Julie also suggested to Kelly that she needed a more powerful computer, a new microphone that could cope with the background noise in the open plan office and maybe some partitions around Kelly's desk to block the background noise even more. Kelly said that she had previously tried to get a better computer but was told that she was not allowed. Julie had Kelly give her the names and contact details of Kelly's manager, the IT manager and the Occupational Health and Safety manager. Within a month, Julie had negotiated an upgraded

computer for Kelly and arranged for the department to buy Kelly a sophisticated noise canceling headset microphone. Julie said that Kelly's
manager was willing to find some old partitions but Kelly thought that
would segregate her from the office and she didn't want that.

Today, Julie and Kelly are having their fifth meeting. Kelly asks Julie
about the most recent upgrade to the version of the speech recognition
software that she uses. Julie says that she thinks it's a worthwhile upgrade to the current version as it has better integration with some email
programs, but that she wouldn't recommend it for Kelly as some of Julie's
clients are reporting problems integrating it with the particular email program that Kelly's department uses. This sort of problem with the speech
recognition software not properly integrating with the software that a department uses is common. Kelly tends to write reports and send emails
so she doesn't need to have a lot of software integrated with her speech
recognition software and she doesn't like the thought of working in a department where there are a lot of different applications to master by voice.
Even browsing the department intranet is hard enough, and all Kelly does
with that is to look up phone numbers of other people in the department.

Kelly also asks Julie to give her a hand writing a macro for a new style
of report that the department is adopting. A macro is a small program
that lets Kelly speak a simple phrase to make complicated things happen
in different programs, such as word processing or spreadsheets. The last
thing that Kelly and Julie talk about is the additional support that Kelly is
starting to get from the IT section now that there are several other speech
recognition users in Kelly's department. It's still a struggle to have anyone
come and help with what might be a speech recognition problem but one
of the more junior IT support people makes time in his lunch hour to help
out the speech recognition users. Julie suggests that Kelly start a speech
recognition users' group to give the speech users a stronger bargaining position when trying to achieve change. Kelly is a bit shy about approaching
management but she says she will think about it.

Julie leaves and Kelly starts talking to her computer again.

4.3   Comparison of Scenarios

My purpose in this chapter was to present two naturalistic accounts of the use of speech recognition to show that the use, and usefulness, of ASR software is not only located in the design of the user interface but in many areas outside of the user-software pair. The two cases presented here show two different sides of the experience of using ASR software in a productive environment. Paul, the Hansard editor, has a very good experience using the ASR software while Kelly, the middle level manager, initially has a poor experience, but persists with the software because it is better than not working and better than being in physical pain.

Paul's good experience stems from the support he has within his organisation, the fit between the work practice that Hansard uses and the way the software is used within that work practice and the fact that the work environment is amenable to the ASR software. Kelly's poorer experience shows a lack of support within her organisation, a mis-match between the work practice and the software and a work environment that is not speech-recognition friendly. These themes will be explored further in chapter 6.

The next chapter presents a rich, naturalistic, description of a workplace where ASR could be used—the ACT Magistrates Court.

CHAPTER 5

The Australian Capital Territory Magistrates Court

The previous chapter described two workplaces where automatic speech recognition (ASR) software is already in use. This chapter describes a workplace where ASR could be introduced. Because the contention of this thesis is that the usability of ASR is not solely determined by high recognition accuracy but is heavily dependent on the integration of ASR systems with work processes and work organisation it is useful to examine the work process of a workplace where ASR software is not yet used. That workplace is the ACT Magistrates Court.

My association with the Court arose through my association with the Court of the Future project (Court of the Future website, 2004, 2004) at the University of Canberra. We were approached by the Chief Magistrate of the Court to investigate the introduction of ASR technology to the courtroom for use by the magistrate in the process of recording outcomes, which is a highly charged moment in the Court when a magistrate speaks an outcome for the case that he or she is hearing. An outcome may be a sentence, for example a fine or jail term or it may be the decision to set a case over to allow all the parties to the case more time to gather relevant information. An outcome may also be a procedural decision specific to the Court such as a request by the magistrate for any number of specialised reports that are used to inform the actual sentence when it is finally delivered. The question we were attempting to answer was: what form would ASR technology take at the Court?

After some preliminary ethnographic work at the Court it emerged that the magistrate's act of speaking an outcome was not an event that was self contained but the beginning of a process distributed in space and time throughout the Court. This chapter describes the deeper ethnographic investigation of the processes involved in determining and recording outcomes of cases and shows that the process involves many different court workers, each performing detailed work that contributes to the final outcome (Kraal et al., 2004).

In this chapter I will describe the work situations of the people involved in the process of determining outcomes in the Court. The main human characters are Rob Cowley, the magistrate; Clare, Rob's associate; Brenda, the list clerk; Peta, the monitor; Carmen, who works in the "after court" section; and Dana and Frances, two bail office workers. This chapter also includes a description of the work of a non-human character, the *defendant's folder*.

Unlike the previous chapter on the Hansard department at Parliament House and ASR users in the Public Service, the scenarios and characters in this section are based on real people and their activities and are not amalgams from many interviews. Names have been changed where appropriate and all care has been taken to ensure that the scenarios and descriptions are respectful of the people and their work. Whenever reference is made to a court room, that court room is Court Room One (see figure 5.1, 75).

5.1   Communicating Outcomes Defined

Communicating outcomes of cases is a complicated process that involves many court workers whose work in the process is distributed in space and time. As the work process moves through the Court and each worker performs their specific task, several documents accompany the process: the bench sheet and other documents in a manila folder, each folder belong-

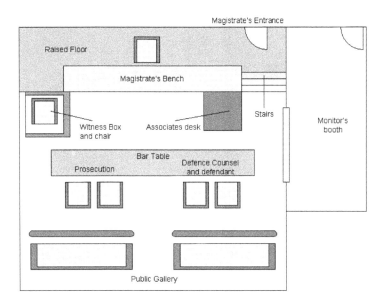

Figure 5.1: Plan Diagram of Court Room One at the ACT Magistrates Court. Note: not to scale.

ing to one defendant (hereafter the defendant's folder), and the monitor's notes. A monitor is a person who, in the ACT Magistrates Court, listens to one or more courts in session from a special booth within the Court building and annotates the audio recording of the particular court(s) in session (for more on the monitor, see section 5.2.4).

The flow chart below shows which court workers are involved, and where the documents are, in the various parts of the process. The flow chart only deals with the "A-list". The A-list is the busiest part of the day in a court room and involves the greatest number of people in the shortest amount of time appearing before the magistrate. The A-list is reserved for the first appearance of people who have been summonsed to court. Most of the work done in an A-list session is procedural and final judgments are not made unless the magistrate in charge thinks that the case can be heard

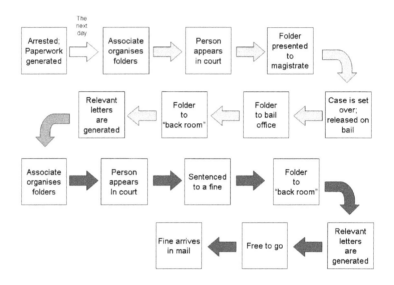

Figure 5.2: An example time-line for a defendant's appearance in court.

in a short amount of time. This means that people who are pleading guilty to minor charges are dealt with almost immediately but people who are pleading not guilty are re-scheduled for a later time.

The process of sentencing begins when a person is arrested. They may be kept in the lock-up overnight or they may be released and given a court date. If their court appearance is adjourned they may be ordered to reappear at court on a subsequent date, or they may be given bail, which obliges them to reappear on a certain date.

The paperwork that drives the process of sentencing begins at a police station but is collated and takes the form that this thesis is concerned with in the court building itself.

5.2 Court Workers

This section tells the story of the basic work process of sentencing at the ACT Magistrates Court through the eyes of several Court workers. These descriptions of court workers begin with defendant's folder, a non-human actor in the process. The defendant's folder has a coordinating role in the process. The human court workers are then presented. Each court worker is presented in sequence, or as close to in sequence as possible so that the reader can place them into a coherent whole and the process can be made clear.

DRAMATIS PERSONAE

The Defendant's Folder A non-human actor

Clare The Magistrate's Associate

Mr Rob Cowley The Magistrate

Peta The Monitor

Brenda The List Clerk

Carmen Works in the "After Court" section

Dana Works in the Bail Office

Frances The more senior court worker in the Bail Office

Amanda The Corrective Services officer assigned to the Bail Office

Figure 5.3: Dramatis Personae: Actors in the Performance of Communicating Outcomes in the Court.

5.2.1   The Defendant's Folder

The defendant's folder is a coloured manila folder affixed with a label with a case number, the defendant's name and the *informant*'s name. The informant is generally the police officer who made the arrest and caused the defendant to have to appear in Court. A defendant will typically have one folder in use at any one time and efforts are made to ensure that new defendant+informant pairs are not created while a defendant's case is being heard. If it is at all possible, each folder is assigned to one magistrate to ensure continuity.

A new folder, representing a new case before the Court, does not contain much apart from an arrest report and associated documents. As a defendant moves through the system of appearances in court, the folder acquires more paper and more documents that represent processes and outcomes of processes that have been ordered by the magistrate in charge of the case. The most important document in a folder is the bench sheet.

The bench sheet is actually a series of notes that the magistrate makes as a case unfolds in front of him or her. The bench sheet also holds the written record of the magistrate's decision at the adjournment of each case. Bench sheets begin as blank pieces of A4 paper. Each magistrate uses their bench sheet in slightly different ways, though some date theirs and others rely on their associates to do so. As court unfolds, a bench sheet is covered with the magistrate's sparse handwritten notes on the case. The notes are sparse because the lawyers in the case typically provide documents detailing their positions which are incorporated into the folder.

The most important information that the bench sheet holds is the magistrate's decision on the most recent outcome of the case. The outcome can be recorded in handwriting or by a rubber stamp (see figures 5.4, 5.5 and 5.6, 79) which may or may not require extra handwritten detail to be provided. The stamps are useful for oft-repeated decisions or parts of decisions but a lot of the detail of an outcome ends up being handwritten because it is unique to each case.

As part of a decision for an outcome a magistrate may request one of a number of reports to be produced by a number of agencies. The reports

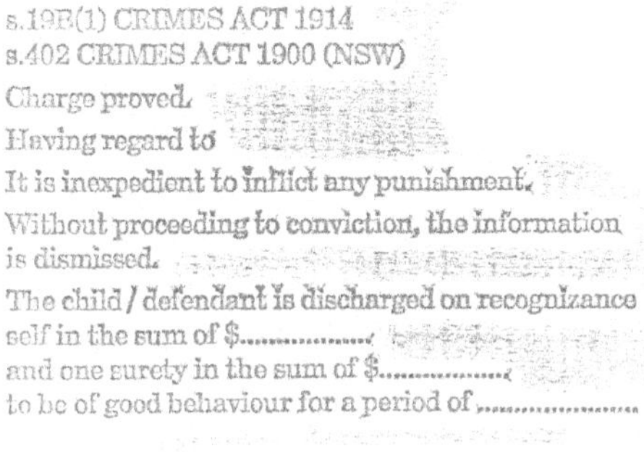

s.19B(1) CRIMES ACT 1914
s.402 CRIMES ACT 1900 (NSW)
Charge proved.
Having regard to
It is inexpedient to inflict any punishment.
Without proceeding to conviction, the information
is dismissed.
The child / defendant is discharged on recognizance
self in the sum of $.................
and one surety in the sum of $.................
to be of good behaviour for a period of ....

Figure 5.4: A typical stamp used by a magistrate.

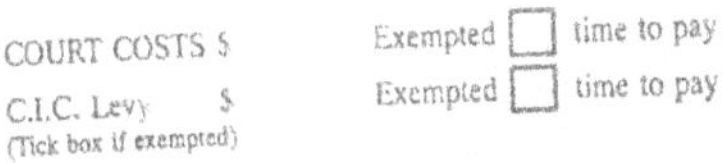

COURT COSTS $ Exempted [ ] time to pay
C.I.C. Levy $ Exempted [ ] time to pay
(Tick box if exempted)

Figure 5.5: This stamp is applied to many bench sheets.

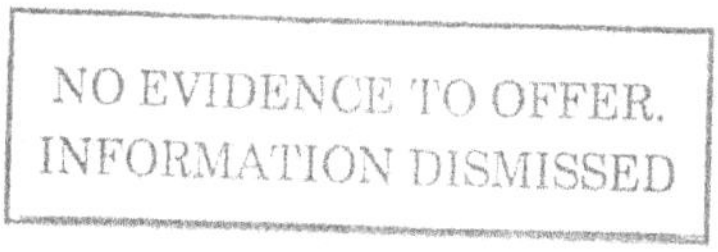

NO EVIDENCE TO OFFER.
INFORMATION DISMISSED

Figure 5.6: This stamp is used when the magistrate decides there is no case
to answer.

can encompass (but are not limited to) a defendant's mental health, drug or alcohol dependency and so on. These reports find their way into the folder in a timely manner and are available to the magistrate the next time a defendant appears in court.

The folder allows different workers in the Court to coordinate their work. As the folders move from storage to the Magistrate's Associate, the Magistrate, back to the Associate, perhaps to the Bail Office and then to the After Court section the folder and the information it contains are used to communicate decisions and represent the work done in court to those who could not be there. The folder even acts as a window into the semi-distant past as it is used by magistrates *now* to show what happened in court *then*.

Making the folders useful in the Court requires a lot of effort on the part of people using them but because the folders are so useful and essential to their work, the maintenance of the folders is not arduous. The folders are so useful that they have their own room in the court and a person who is dedicated to their storage. Folders are stored in the file room and are filed according to case number. The file room is large and well-lit, if slightly out of the way. Many workers in the court use defendants' folders in their job, but filing them after use and finding them when someone wants them is the work of the File Clerk. Each day before Court (or on the afternoon of the previous day) the File Clerk gets the files out for each Court session and puts them in a pigeon hole for the Magistrate's Associate.

5.2.2   The Magistrate's Associate

Clare is the Chief Magistrate's Associate. Her job is to prepare all the documents that the Magistrate may need in Court and to assist him in managing the courtroom itself.

Clare's day starts when she has a coffee with Charlie, the head custodial officer. Clare has worked at the Court for almost 18 years and she knows everyone. After her coffee with Charlie, she walks to the file room and picks up the large stack of files that Terry, the file clerk, has placed in her pigeon hole. Organising the files will be the bulk of Clare's work for the morning. She carries the files back to the lift and goes up a few floors to her office. Clare's office is small, but because she is the Chief Magistrate's associate, it's very nice, with a large window over well-kept gardens. Clare has personalised her office with pictures of her children and trinkets from her trips overseas.

Clare sits at her desk and with the files beside her computer, calls up a screen in the Court's computer system that will let her check that she has all the files she needs for the A-list tomorrow. The first step is just to go through the stack of files and check that they're all there. The list is a little shorter than usual today but there's still a lot to do. A few of the files listed on the screen have electronic notes appended to them that Clare checks. Sometimes the notes are simple but often they relate to procedural matters such as a folder relating to a person who is "in custody" in the cells below the Court which means that Clare has to deal with the folder (and its associated files) differently to the other folders.

As she checks that she does indeed have each folder listed, Clare re-arranges the contents to make things easier for Mr Cowley when he's on the bench. Clare is very familiar with every sort of document that could appear in a folder and does this quickly. The document that goes on the top of the file is called an Affidavit of Service and always comes with two copies. The Affidavit of Service describes how a defendant came to be in court. Clare used to know why there were two copies but it has become part of the routine to separate the copies and leave one on top of the file. At least there are only two. In some places in the Court, some documents have to be printed with up to five copies. As well as putting an Affidavit of Service on top of the file, Clare writes the date on the reverse and, if it's on the list on the computer, also adds the name of the prosecutor. Some of the other associates have started to use stamps for the date, but Clare and Mr Cowley have their routine and Clare doesn't think that a stamp would make a difference to how she organised these files.

Clare moves through the next few files, check, open, reorder, close. This next file is quite thick. The first charge is for what's called "assault police" which is what it sounds like. A few months ago something at the police's end of this process must have changed because the notes in the file used to just say "assault police" but now there's a story that goes for half a page on what the arresting officer said and what the defendant said. The affidavit goes on top and this defendant has been bailed so Clare staples his bail documents to the inside cover of the folder so that the Mr Cowley can see them quickly. All of the associates prepare their files this way but, as with the date stamp, the longer you work for one magistrate the more the routine changes.

This next folder isn't in the list on the screen so Clare adds it to the list by entering its file number. Reorder, staple, close. Next.

A few more routine folders and then one that contains files for a matter that's still current. Mr Ridgeway, one of the other magistrates, is still dealing with this case, so Clare transfers it to his list by pressing a few keys. The folder itself goes on the floor in the "back to the file room" pile.

Clare recognises the next name on the computer screen. He's been in and out of court since he was about 12. He keeps disappearing when he's on bail so he just gets more and more arrests from different officers which means more and more folders. Clare has five folders for this fellow in front of her and she has to go through all of them looking for the right affidavits and bail documents. He's on his second appearance for this set of charges and Mr Cowley has ordered a pre-sentence report and a CADAS (Court Alcohol and Drug Assessment Service) report. The pre-sentence report might be in a box downstairs but judging by the date it was requested it's possible that it will be hand delivered to court tomorrow morning, along with the CADAS report.

The next few names are also familiar to Clare. Some of them she feels sorry for but some are, in her opinion, quite disturbed.

This next name doesn't have a folder in Clare's pile. Maybe it was transferred into her list by one of the other associates after Terry pulled the folders from the shelves downstairs. Clare is senior enough in the Court to call down to Terry and have him bring the file to her.

A few more very routine folders and Clare is ready for the A-list session with Mr Cowley. She takes the big pile of sorted folders in her arms and makes her way to the lift at the end of the corridor. Bobby, a new secretary, has just gotten out of the lift. Clare says good morning to Bobby, gets in

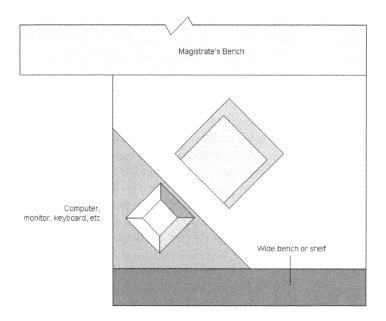

Figure 5.7: Plan view of Claire's desk inside the Court.

the lift and makes her way down to Court Room One, where the A-list will be held today.

Entering the courtroom from the back door (only the public enter from the front) Clare sits down at her little desk (see figure 5.7) in a special booth, down from and to the left of the Magistrate's Bench. She sets down the folders on the wide bench that doubles as her shelf, logs in to her computer and waits for Mr Cowley, who's always late.

5.2.3 The Magistrate

Rob Cowley is the Chief Magistrate. Being Chief Magistrate is one of many hats he wears and although he's always busy he manages to come across as relaxed. People who work in the ACT Magistrates Court address Rob as Mr Cowley out of respect. He's Rob to everyone else except when he's on the bench where he's Mr Cowley to everyone. This description, though written in third person, is Rob's story so he is referred to by his first name.

It's 9.30am on a Friday and it's time for the A-list. The A-list can seem like rough justice because it's fast paced but Rob tries to keep the court calm and usually manages. At 9.35am Rob enters the courtroom in his black robe and takes his seat behind the bench. The court is quite modern with pale coloured wood panels and desks. As soon as Rob sits down, Clare, who sits below him and to his left, hands him the folder of the first defendant. Rob speaks his name "Dennis Horton?" and Mr Horton steps forward along with his lawyer. A brief exchange between Rob and Mr Horton ensues, followed by a statement by the public prosecutor. Rob decides that Mr Horton's case is fairly simple and says "Pursuant to section 402 of the Crimes Act, having regard to character and circumstances, without conviction the defendant will be discharged on recognisance self of $2000 to be of good behaviour for six months. Included in that are court costs of $52 and the CIC levy of $50". He then selects one of the large rubber stamps on the bench, applies it to the bench sheet in front of him and fills in the details of Mr Horton's release, recognisance amount and the court costs and CIC (Criminal Injuries Compensation scheme) levy. Rob hands the folder to Clare in exchange for the next one in the list. Mr Horton leaves the court looking quite relieved that the experience is over.

More cases move through the Court. Rob deals with them calmly, though quickly. Many cases are put on hold for later times to allow defendants, or more usually defendants' lawyers, time to prepare their case. Rob consults with Brenda, the list clerk, as to when he can schedule the cases he's setting down for future dates.

The Court breaks for fifteen minutes at 10.30am. Rob, Brenda and Clare leave through a door at the rear of the Court. Most of the public gallery stay where they are, though some stand to stretch. If you look around carefully, it's possible to guess who is representing themselves, who is an old hand at court and who is nervous because it's their first time.

10.45am and court resumes. The case is one of shop lifting and the defendant, a Mr Geoff Bluestone pleads guilty. The police have raised charges for each of the twelve DVDs stolen as well as for resisting arrest. It turns out that while Mr Bluestone was waiting for his court date he stole five DVDs from a different shop and the police have charged him for each item stolen in that instance as well. Rob understands why the police raise the charges in that way, but he finds that it greatly slows him down when he's pronouncing sentence so before he begins he waives eleven of the first set of theft charges and four of the second set. He asks if the public prosecutor and Mr Bluestone's lawyer have any problems with that before formally announcing his decision. Rob also notes the waived charges on the bench sheet by listing the charge numbers. It's important for the people who do the after-court work to know which charges are actually being dealt with.

After a statement by the public prosecutor and the defendant's lawyer, Rob asks Mr Bluestone if he has anything to say. Mr Bluestone says that he's sorry for what he's done and that his drug habit forces him to steal things. Rob says that Mr Bluestone has appeared before and this is the last time that he can be let off without imprisonment. When Rob addresses the lawyers and the defendant he takes a professional tone but one that also has a degree of intimacy about it - he is only addressing the three people in front of him. When he comes to pronounce sentence, Rob's tone of voice changes slightly as he addresses the entire court:

"The formal sentence of the court will be that in each case he will be convicted without passing sentence and released on recognisance self $1000 to be of good behaviour for two years. I will impose no additional punishment on that."

As Rob speaks he finds the correct stamp and applies it to the two charges of theft. He finishes speaking and fills in the details.

"In regard to the assault police, he'll be convicted and sentenced to six months imprisonment. I order he be released forthwith on recognisance self $1000 to be of good behaviour for two years. I'm going to order that for a period of 12 months he undergoes supervision on probation including any direction for treatment, counseling, residence breath testing, and urine analysis. I'm also going to order that for a period of six months he accept CADAS supervision and also not to consume any alcohol or drugs except by medical prescription."

Rob speaks to Mr Bluestone again, asking him if he understands the requirements set out and the implications of not attending the CADAS supervision. Mr Bluestone indicates that he does.

Rob uses the same stamp for the assault police charge as the theft charges and fills it in but writes all the requirements of the 12 months' supervision and CADAS supervision by hand. By this late in the court session Rob's done a lot of writing and his penmanship has deteriorated. Including the stamp the details for this sentence fill most of the page.

Court progresses through a series of other theft charges and moves on to traffic matters. Some traffic matters are simple drink driving charges where Rob gives the defendant a stern talking to and the legislated fine. Others are speeding tickets. Almost everyone pleads guilty. The most complex traffic matters occur when the blood-alcohol reading from a drink driving charge is so high that Rob is compelled to suspend the defendant's licence. Ms Daly is the most complex of these cases in this session of the A-list. Rob quickly moves through the routine part of sentencing Ms Daly, informing her that he will suspend her licence for a period of 6 months. Ms Daly's lawyer speaks up to tell Rob that Ms Daly requires her licence as she is a part-time student as well as a shift worker. Rob asks Ms Daly what times and days of the week she studies and works as well as where she lives and the location of her college and workplace. Rob is familiar enough with the part of Canberra that Ms Daly lives in that he suggest that she catches the bus to college and even names the route number. He concedes that she will probably need her car to get to and from work and begins the process of filling out a restricted licence form. The form requires detail of the level of hours between which Ms Daly is allowed to drive. Her restricted licence will have that information on the reverse. Negotiating with Ms Daly when she works is more of a task for Rob than filling the form, but only just. It's pretty clear that Ms Daly would rather have a non-restricted licence.

Court continues until just after 1pm. Everyone seems a bit the worse for wear when the last case is finished.

5.2.4   The Monitor

Peta is a monitor at the ACT Magistrates Court. She's studying law part-time and works at the court part-time.

Peta's job is to listen to what happens in court and note down specific things the presiding magistrate says. Peta works in a small booth (see figure 5.8) between the walls of the courtrooms. The booth looks in to two court rooms, one on each side, through one-way glass. The booth also has seats, computers and closed circuit monitors for two people. Today there is no-one else in the booth except Peta.

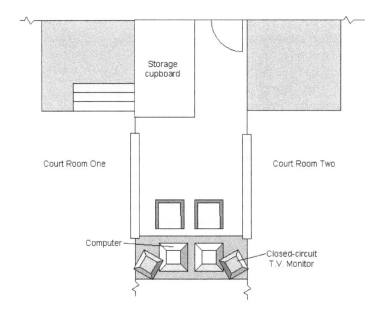

Figure 5.8: Plan view of the Monitor's Booth between Court Room One and Two. Note: not to scale.

The Chief Magistrate is presiding over the A-list which means that Peta will be quite busy for the next four hours. A regular court session with one long case is much quieter than the A-list with its many short cases and sometimes Peta can get some study done.

Peta has two things to do, though both are related. She has a list of the charges that the court will be hearing today, ordered by defendant. Today's list is about 10 pages long. Each charge has a box next to it for Peta to write in. Peta's job is to mark on the list which charges are being heard and which the magistrate is dismissing. On the ones that are being heard, Peta writes on the sheet, very briefly, the outcome, whether it's bail, a jail sentence, a fine or something else.

Peta also uses a computer with special software to annotate an audio recording of what's happening in court. The software allows Peta to make time-aligned notes that are linked to the audio recording. In non-A-list court, Peta annotates the recording every time anyone, magistrates, defendants, witnesses, lawyers, speaks in court. Because this is the A-list, all Peta has to do is use the computer program to record who the lawyers are when each new defendant is called.

Everything is fairly routine until a Mr Bluestone is called. Peta notes Mr Bluestone's lawyer's name in the computer.

Mr Bluestone's list of charges goes for several pages in Peta's list. Peta has been a monitor for long enough to expect the Chief Magistrate to waive most of the charges early in the hearing so she listens with her pen poised over the sheet. As the Magistrate says what charges he'll be waiving Peta crosses them out, leaving her with three remaining, spread over two pages. Peta listens to the proceedings but only half-heartedly as it's pretty clear to her where this is going. Sure enough, the Magistrate suspends the sentences and proscribes supervision on probation. Using a set of abbreviations Peta notes down the outcome.

The next defendant is a Mr Day. He pleads not guilty to a speeding ticket, a fairly unusual occurrence. The Magistrate asks the public prosecutor how many witnesses he will want to call and does the same for Mr Day's lawyer. Peta can see Brenda step forward and have a brief conversation with the Magistrate and the Associate. Brenda steps back and the Magistrate says for Mr Day to reappear in two week's time. Peta notes down the date on her sheet.

5.2.5 The List Clerk

Brenda is the List Clerk at the ACT Magistrates Court. It's Brenda's job to ensure that the court dates are filled as efficiently as possible.

The first part of Brenda's day takes part in court for the A-list where she assists the presiding magistrate in scheduling a range of matters such as bail variations, pleas of not guilty and criminal traffic hearings. There are also some matters that Brenda has no input into such as sentencing hearings that are usually so quick that the magistrates are allowed to squeeze them in around other scheduled matters.

In the A-list today, the most difficult task Brenda had was to schedule a plea of not guilty for a Mr Day. There seems to have been a lot of people caught for speeding lately and the rosters that Brenda takes into court with her were very full. Fortunately, the lawyers in Mr Day's case seem content with calling only one witness each so Brenda was able to make a note in the roster and schedule a 20 minute hearing for Mr Day in two week's time, rather than in three months.

The second part of Brenda's day comes when she receives the list from the monitor. Brenda uses the monitor's list to construct the rosters for the hearings that she provided dates for in the A-list. The computer helps a little but it's still a job of shuffling appointments around and judgment calls.

5.2.6 The After Court

Carmen works in the After Court section of the ACT Magistrates Court. Her job is to update the computer system with all of the day's happenings in court. Carmen works mainly from the bench sheets that the magistrates make their notes on during court.

Today the files that Carmen has to work from are a big pile of manila folders, mainly from the previous day's A-list though there are some regular court folders as well. Carmen thinks of her job as "bulk work". She opens each folder and checks it against the relevant entry on the computer system. By looking at the notes made by the magistrate she is able to update the computer system; the magistrates work from paper but the court

is run by the computer. Carmen has become something of an expert in
the different magistrates' handwriting, particularly Rob's but Mr Leon's
occasionally gives her a bit of trouble. Fortunately Cherie, who works at
the next desk, seems to be able to read Mr Leon's writing so Carmen will
occasionally ask her for help. Sometimes Carmen will have to go to Mr
Leon, or one of the other magistrates to verify what they've written.

Reading the day's folders isn't that difficult, once you've got the knack,
Carmen tells the new people in the After Court section, but doing the interpretation takes a little longer to learn. A lot of the extra processing
involves interpreting one line of spidery handwriting and knowing that
"$50 fine on all charges" can sometimes mean thirty different entries into
the computer.

Using the Court's computer system is an exercise in remembering lots
of letter-and-number codes. For each part of a sentence, Carmen has to
enter a code which could look like "CDF54" or "BAIL" or a wide range of
similar combinations. There's also a special code to say that a particular
case is finalised. Depending on the charges and what the magistrate has
ordered there can be seven or eight codes for one entry, though less than
five is more usual.

Most of the codes go into the system so that Carmen can prepare various letters that are sent out to defendants. Some letters, like adjournment
letters, are generated by the computer system but others Carmen has to
create individually. Doing the automatically generated letters is a matter
of selecting the right menu option in the computer system at the end of a
day's processing. The printed letters come out of the printer and are then
mailed to the address printed on the letter. The individually created letters take much longer. Carmen has to know when a letter is required and
then has to put the correct information into a special template letter she
has created in Microsoft Word. It's a fairly involved process that means
she has to do lots of flipping back and forth between the Court's computer
system and the letter.

Some of the letters that Carmen has to prepare herself are a way of
working around the computer system. Because of the way the system
works, certain combinations of codes for the one case can produce two

letters which is confusing for everyone concerned, particularly the defendants. Carmen knows which codes these are and in which situations they will produce two letters so sometimes she tells the system not to print a letter and creates a letter herself. One of the more common times this happens is when a defendant is not given a formal sentence but is made to pay court costs and the CIC levy, a combination of codes that makes the system produce two letters.

If the defendant is sentenced to a fine and is made to pay court costs and the CIC, only one letter is produced. Fortunately most of the time the system gets it right and only produces one letter.

When the magistrate and the lawyers for a particular case decide that they'll amend the charges, or dismiss some, Carmen has to make those changes in the system before she can begin entering the codes.

5.2.7 The Bail Office

Dana and Frances work in the Bail Office at the ACT Magistrates Court. It's their job to prepare the bail documents for people who leave the court on bail. The bail documents are prepared as people leave the courtroom.

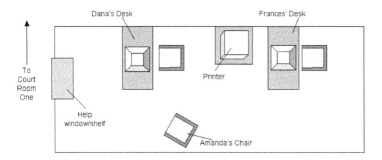

Figure 5.9: Plan view of the Bail Office. Note: Not to scale.

The bail office (see figure 5.9) is a room just off the small corridor that leads to the courtroom. The door to the bail office is cut in half so that the top half can be opened when the bottom is closed, turning the doorway into a sort of counter. Across the corridor from the bail office is a waiting room.

When the day starts Dana and Frances organise the bail office by tidying it up a little, and by working through some bail-related documents that have come in with people who were arrested overnight. They do this together with Dana reading out the details to Frances who is logged into the Court's computer system. It doesn't take that long and while they wait for the first person to emerge from the court they talk about their respective evenings. Dana watched *Shrek* with her boyfriend and Frances ate at a new restaurant with her friends. The conversation wanders to other movies they have seen and other restaurants that are worth visiting until a woman is escorted from the court by a bailiff. The bailiff carries the defendant's folder with the newly amended bench sheet and other documents.

The bailiff tells the woman to sit in the waiting room and hands the folder across the counter to Frances. The folder contains the bench sheet and other documents. Frances takes it to Dana and reads the relevant defendant's name, the charge number and the bail conditions to her as Dana types. Frances crosses this defendant off a list of everyone who is scheduled to appear in court today. Dana finishes typing and prints the bail order in triplicate by specifying "3" in the number-of-copies-field of the print window. The bail orders are always printed in triplicate.

Frances collects the printouts and calls the woman's name, "Mrs Franklin!". Mrs Franklin approaches and Frances tells her that this is a bail document that means that she has to come back to court or a warrant will be issued for her arrest and she will have to pay the bail amount. Frances asks if she understands and Mrs Franklin nods her head. Frances has her sign all three copies of the printouts and then they tell her she's free to go and to come back in two weeks, as the bail order says. Frances files the three forms in three different boxes for collection by people who work in different parts of the Court.

It's a pretty slow day today and the bail office is pretty quiet. Sometimes the waiting room gets quite full and Frances is at the counter having defendants sign documents while Dana puzzles out the bail orders for herself. Every magistrate has a different way of writing their bail orders, from tick-a-box forms to abbreviated handwriting to full sentences. The hardest ones to understand are the abbreviated ones. Sometimes Dana has to go into the court and wait for a good time to interrupt so the magistrate can clarify what they meant on the bench sheet.

The bail office part of the computer system works by codes, much like the one Carmen uses in the after court section. There are some procedures that they do in the bail office that the computer system hasn't caught up with and, like Carmen, Dana and Frances have a set of templates in Microsoft Word that they use to create the appropriate letters. One of the templates that Dana created is the *Warrant of removal to an approved mental health facility* template which is used when a magistrate sends someone to a hospital for evaluation. In the computer system Dana has to make a note that the person is being sent to a mental health facility in a special "note" field and then use the template to create a letter that will be sent around the Court and to other places.

Amanda comes into the bail office. She works in the corrective services department and she speaks to people who have been ordered to be "under supervision" and lets them know what that means. Frances tells her it's a pretty quiet day and she doesn't think that Amanda will have a lot to do.

A man approaches the counter and asks Frances if his cousin, Mr Beale is in this court today. Frances checks the list on Dana's desk and says that Mr Beale has already appeared. The man says, "thanks" and hurries away.

Some more defendants come out and Dana and Frances process their bail orders. In between, Amanda and Frances joke that everyone must have behaved themselves last night.

The next defendant, a Ms Tower, has been given a Mental Health Order by the magistrate. Dana thinks it's a good thing that it's quiet today because MHOs take a long time. Dana has to enter Ms Tower's information into the computer system, use a template to create a letter and make several phone calls to various parts of the Court to organise the right sort of supervision for Ms Tower.

When court finally finishes, Dana and Frances are glad to leave the bail office and go back to their regular desks.

5.3 Summary

The work of the ACT Magistrates Court is complex. The descriptions, above, have focused on a small part of the work that relates to the process of communicating sentences. These ethnographic descriptions are used in chapter 7 where the work process is analysed and an ASR system is proposed to support communicating sentences.

The next chapter revisits the use of ASR in the Australian Public Service by injured users and in the Hansard department. The use of ASR is analysed and the properties that make it useful and usable in the workplace are examined.

CHAPTER 6

Analysis of Automatic Speech Recognition Ethnography

In this chapter, the interviews with automatic speech recognition (ASR) application users are analysed and concepts that emerge from the analysis discussed.

The interviews analysed in this chapter were conducted with eight ASR application users. The interviewees were six ASR application users who had previously suffered from a "workplace overuse injury" who all worked in different agencies within the Australian Public Service and two ASR application users who worked in the Hansard department of Parliament House.

The analysis of the interviews shows that using ASR dictation systems in the workplace is difficult, though not for the reasons typically reported in the recent ASR usability literature (see section 2.2). The next section briefly recounts some of the findings of usability studies of ASR systems and compares the program of action inscribed in the software with the program of action that the interviewees found necessary to make the software useful.

6.1   The Problem with Automatic Speech Recognition Applications

Automatic speech recognition is marketed as an easy-to-use system that allows a person to control a computer only by voice.

> "Does creating documents, spreadsheets and e-mail take up a large part of your day? Work faster and more productively with Dragon NaturallySpeaking Preferred speech recognition software!" (Dragon NaturallySpeaking Preferred 6, 2002)

When a person is introduced to an ASR application software package it can be for a number of reasons. In my research, the situations are the occurrence of an occupational overuse injury or the occasion of starting work as a Hansard editor. In both cases, the software application that is used is the same, however, the overuse injury sufferers and the Hansard editors have different experiences.

The Hansard editors do not find the application hard to integrate into their work while the Public Service users find the integration difficult. By comparing the two groups of users, I will show that the usefulness of ASR application software is not only a part of the software application itself but is also bound up in the social situation in which it is used.

Use of ASR application software is often seen as a single user talking to a computer, speaking sentences to compose a letter or a report. While it is true that one user's "work" may involve writing a letter, the work involved in actually being able to compose that letter with an ASR application is not only in using the application but also in integrating the application into a much wider process. For example, the letter may be in reply to a phone call which followed up a report that showed that a department's budget was going to be 80% lower in the next financial year. The letter does not exist in isolation but as part of a larger process of work.

When examining the usability and utility of ASR applications, the recognition rate of the software is a dominant focus (Huang et al., 1999). The recognition rate is a measure of the ability of the software to recognise the words spoken by the user. The goal is always to have a higher recognition rate. Most new commercial ASR software applications, for example Dragon NaturallySpeaking, are advertised as having a high (perhaps 98%) recognition rate. In bad situations, such as noisy rooms, or in very difficult situations, for example in broadcast speech recognition, the recognition rate can be as low as 40% – only four words in ten correct (Prasad et al., 2002).

I believe that the focus on recognition accuracy produces the narrow view that many people have of using ASR applications which only involves recognising words as they are spoken. For many people this matches their experience of taking dictation or talking to an intelligent secretary when dictating a letter. That process takes the form of calling the secretary into the room and informing them that dictation will begin. Dictation can proceed in fits and starts and the speaker can revise at any time. If the speaker stumbles over a word, or starts a sentence badly, they are able to revise as they speak, using commonly accepted conversational methods. The speaker and the secretary cooperate to make dictation work.

Dictating to an ASR software package is quite a different task. After starting the software, and assuming that the software is trained to the speaker's voice, the computer records every word spoken. If the user misspeaks, stumbles or changes their mind in the middle of a sentence, the computer does not know but treats all sounds as intended and attempts to recognise each sound as a speak act. The resulting text does not look as if it was taken by a person. Even if the recognition was perfect, the dictation would contain every stumble and misspoken word. At the current state of the art the software application cannot cooperate with the user in the same way as the secretary does with the speaker.

Every person who uses an ASR application productively overcomes the deficiencies in the application by changing how they work, changing their work environment and coming to understand, at least a little, how the ASR application works. The user must make up for the lack of cooperation from the software.

The idea that using a software application in a particular environment involves some degree of modification to the environment, the existing work practices and even to users themselves is not a new one, indeed it is almost a cliché to suggest that use is contextual. It is equally trite to say that ASR software is hard to use. This chapter is not about what ASR systems lack that makes them hard to use or the fact that the use of ASR systems is situated. This chapter is about how people who use ASR systems overcome the deficiencies in the software and, through their efforts, make the software usable. The next sections will show how these users manage their use of ASR application software in the complex social environment in which they find themselves—the workplace.

The next sections use the lenses of the *Locales Framework* (Fitzpatrick, 2003) and Actor-Network Theory (Callon, 1986b; Latour, 1987; Law, 2003) to show how ASR systems are made useful in the workplace. The task of using an ASR system in the workplace is not a simple one and involves many heterogeneous actors, both human and non-human, social and technical. The trajectory that each user follows as they make the software useful is necessarily unique. The similarities between the trajectories reveal what is necessary to make ASR dictation software useful in the workplace. The trajectory emerged through the analysis of the interviews with the ASR users. The stories of each user, though different, showed that they had all gone through similar experiences to arrive at successful use of ASR software in their work. The trajectory that each user follows is to first decide to use the software, to leap into using it and then to sustain their use. The next sections explain the trajectory.

6.2   Deciding to Use

This section describes the elements that prompt a person to decide to use ASR software in their work. These elements are:

- Having a reason to use ASR software;

- Having previous experience with similar tools; and,

- Having enough "IT savvy" to be willing to learn a new, and very different, piece of software.

6.2.1 Reasons for Use

There are many reasons for a person to use ASR dictation software in their work, however the most important is that it allows them to do something that was not possible before. In the two case studies in this research, the injured users could not type, and so could not work at all, and dictation software allowed them to return to work. The Hansard department needed a way for its employees to turn words spoken in Parliament into text on a screen. Automatic speech recognition, according to the manager of the Hansard department, is a cost-effective way to achieve that outcome.

The injured interviewees had different specific reasons for using ASR applications. For some, ASR applications were initially seen as a cure-all for the difficulties they have in using a desktop computer. For others, they were told by their health-care provider that using ASR software is the only way that they will be able to return to work. One interviewee said that she came across the software in a magazine article and suggested using it to her superiors and was quite surprised when they agreed.

> "I had a workplace injury that I got in 1997 and I had heard about voice recognition software. I think I read an article about it in the Bulletin or something like that. We had just got a new boss and I went to him and said, 'you know what would be really useful it would be this'. And he said yes! I thought I was really going to have to argue the case but there were two other officers in the division who had RSI problems and they bought it for three of us."

Even though the software was far from perfect it allowed her to do more than before:

> "At the time we were using Macintosh machines so we were using Power Secretary which was quite clunky but I was very grateful because without that I would have had to take weeks and weeks off work and come back, you know, part-time, graduated return."

Other people had dictation software thrust upon them as part of a stream of interventions into their particular form of overuse injury. The interventions typically took a combination of rest, physiotherapy, ergonomic interventions, painkillers, changing jobs or tasks and more. Automatic speech recognition was generally a last resort. At least one interviewee said that she'd been told to use ASR after many and varied attempts to lessen or cure her injury:

> "But people assume because they have ticked off we've given you this drug, you've had physio, you've tried this, you've done that, here's voice[1]."

Some of the injured interviewees' saw ASR as a preventative measure, both for themselves as it prevented them aggravating their injury and for other people who may be prone to similar injuries. Margaret compared the new frame of mind necessary to use ASR instead of typing with the move dentists made to wearing rubber gloves:

> But that said, the same point may be made about other safety features in other occupational groups. When dentists started using rubber gloves at the beginning of the HIV crisis—you know I remember a time when dentists didn't bother using gloves and then they went through, 'oh, I've got to put the gloves on, it feels different'. But the impetus was there, they knew if they didn't do it they could get HIV or hepatitis B or something."

Other interviewees were even more in favour of ASR software, suggesting that it be used instead of keyboard and mouse for all office workers.

The injured interviewees reason for using ASR was management of their injury. Some viewed this management as prevention of re-injury and others saw it as avoidance of the actions that caused their symptoms.

---

[1] Some of the interviewees said "voice recognition" or "voice" for short when they meant "automatic speech recognition"

The Hansard users had ASR introduced to them as part of the tools they used to do their work. No-one at Hansard was injured. The reason for the use of ASR at Hansard was pragmatic. Unlike other tools that could be used to transform the speech of Parliamentarians into text, for example Computer-Aided Transcription (CAT), ASR dictation did not require highly specialised hardware and could be learnt in a relatively short time. Managers at Hansard told me that CAT requires three years training to become sufficiently proficient to be useful at Hansard where ASR requires six months. Not every Hansard editor uses ASR to turn audio into text. Instead, each editor is able to use any method which allows them to keep up with the pace of the work. Some editors use CAT, some are able to type to keep pace with the audio and many use ASR.

The Hansard users who are ASR users, like the injured users, make a decision to use ASR because it allows them to work in ways that they could not otherwise. The experience of using ASR at Hansard is different to that of the injured users because the *organisation* has made the decision to support them. This difference, an individual choosing to use the software and an organisation adopting the software as part of a standard suite of tools will recur and be expanded upon throughout this analysis.

6.2.2 Previous Experience with Similar Tools

It is possible that the interviewees who had previous experience with similar tools to the ASR software were better able to integrate it into their work practices because of that previous experience. I did not focus on previous experience of similar tools during the interviews, however it would be an interesting subject for future studies.

However, one of the interviewees, Jane, had previous experience with similar tools and work practices to those required by dictation software. Jane said that she found it easy to use the software because she:

> "used to work in a law firm and I used to use a dictaphone and I really enjoyed that way of composing. I was familiar with it and comfortable with it. I think if you haven't done something like that before, I think that would be another thing you would

have to adapt to [...]. It's not public speaking, but it is different
for people to do that."

Jane was readily able to see that ASR needed a different working scheme
than typing. The program of action inscribed in the software was different
to typing or speaking: "it's not public speaking, but it is different".

The other interviewees had not used dictation before beginning to use
ASR in their work. Compared to Jane, who was a "true believer" in dictation software, the other interviewees were less enthusiastic about the use
of dictation software by non-injured people.

Jane said that she believed that the technology should be introduced
to all workplaces as a preventative tool for stopping, or at least greatly
minimising workplace overuse injuries. Jane was the first person I interviewed and I put it to the other injured interviewees that ASR software
could be used in that way. None of the other injured interviewees thought
that ASR software was suitable for non-injured people to use as it was too
difficult, and required too much discipline to persist with it when it was
"disobedient" (Read et al., 2002).

Margaret, who compared using dictation software to dentists using
rubber gloves (see quote on p.g. 100) said:

> "There are disciplines that you've got to follow when you're
> using this. There are inconveniences, there are logistical issues
> and unless you have the impetus there, you are not going to
> use it."

Margaret thought that dictation software was useful for injured people
but that non-injured people would find the impositions of the software
too onerous to make them persist with the inconveniences and logistical
issues that arise when using the software.

While the Hansard users interviewed did not have previous experience
with dictation software or techniques, the Hansard *department* has a great
deal of experience with techniques for turning speech into text. Their ear-
liest technique was shorthand, followed by various keyboard-based methods, then Computer Assisted Transcription (which is still keyboard-based)

and then dictation software. The Hansard department's previous experience at turning speech into text made it relatively easy for it to adapt ASR dictation software to its purposes.

The differences in previous experience between the interviewees, or their organisations, did not prevent any of them from learning to use the dictation software. Even Wendy, who was not a productive user, had a high level of knowledge about dictation software. The different previous experiences of the interviewees did, however, give Jane an advantage over the other interviewees when it came to imagining using dictation software in her work. Because Jane was more readily able to imagine using dictation software in her work, she was better able, at least initially, to inscribe the necessary program of action into her work. Jane's experience of adopting dictation software in her work gave her a more positive view of it than Margaret. Where Jane saw dictation software as being useful for everyone, Margaret was more pessimistic.

6.2.3 IT Savvy

Margaret and Jane's views of the usefulness of dictation software (see section 6.2.2) were coloured by their previous experiences of similar tools and techniques. Margaret and Jane had a slightly different "technological frame"[2] (Orlikowski and Gash, 1994) of dictation software. A person's technological frame is their view of technology that is made up of all their previous experiences with other technologies. Margaret and Jane had differences in their technological frame that led them to view dictation software differently, though their technological frames were sufficiently similar that they could both be productive users of dictation software.

Each person's technological frame is constructed in a complex manner, both through previous experiences in the world and through using

---

[2]It is sometimes necessary when analysing fieldwork data to describe what an person is thinking about or how a person has come to a particular point of view. In usability research a person's point of view is often ascribed to their "mental model" (Norman, 1988). While the term "mental model" originates in psychology this thesis approaches usability and design from a sociological perspective so the term "technological frame" (Orlikowski and Gash, 1994) is used to represent much the same concept.

specific technology. The level of familiarity with other, similar, technologies is influential in the construction of elements of a technological frame. The more comfortable a person is with one technology, the more comfortable they will be with a similar technology. This degree of comfort can be called IT savvy. It emerged that the interviewees had varying levels of self-reported "IT savvy" or level of familiarity with information technology.

Wendy, the least "IT savvy" of the interviewees, said:

> "Someone else came to train me and she was in my department and had an injury and had excellent skills in all sorts of areas. And she said well, your system is not working properly. And you see, this is the thing, there is not much expertise around. So, if you don't have the expertise and as you may have gathered I have no technical genes at all, or at least no skills, and I don't know it's not working properly; I only know I'm nearly going mad. So it was a great relief when someone said 'well no wonder, this isn't working properly".

Wendy was at the mercy of the ASR application because she did not know enough about it to know that it was not working properly.

Olivia, in contrast, was very comfortable in using all aspects of the technology. She told how the department in which she was working had "upgraded Lotus Notes and none of the commands work so I had to start again and rewrite all of the commands". Olivia had to use the "macro" feature of the ASR software to adapt the software to using Lotus Notes For Email, an email program that the speech software does not support by default. The ease with which Olivia spoke of having to re-create custom commands that she had already written reveals a level of familiarity with computers in general that Wendy did not possess. Jane, similarly, was quite comfortable writing her own macros and said that she was trying to learn to write macros that would let her create boiler-plate text so that she could say a few words and have a large number of words appear on the screen in an effort to make the ASR application that she used "more efficient than typing things in".

Robyn, the Hansard editor, was very comfortable with computers. In the interview I asked if, when she applied to be an Editor at Hansard, she was aware of the use of ASR software. She said that she was aware when she applied however that she was not really sure what the software was. She was not worried about it because, "I haven't met any software that I couldn't cope with". Robyn had also worked as "a consultant introducing certain software systems", which was a further indication of her level of familiarity with computers.

Yvonne, who worked as an ASR application trainer as well as being a user of the software, was quite sure that users of ASR application software needed to be very familiar with computers in general to be able to best use the software. Speaking of her experience as a trainer, she said,

> "If the underlying platform changes you might have macros that are based on keyboard shortcuts that might change if you get a new version of Word. I think you need to be more computer savvy than if you weren't using Dragon so IT support is really crucial. Even things like backing up your voice files are things most people don't... the average user in most areas that I've worked in doesn't have the skills to do that unless someone shows them a few times."

The interviewees all had differing levels of familiarity with computers and differing levels of comfort in making the computers do their bidding. Their different levels of IT savvy influenced their ability to re-imagine the dictation software as something that could assist them in their work. Olivia and Jane were both very comfortable with writing macros to change and augment the behavior of their dictation software. They were able to imagine new programs of action and then inscribe them in the software, literally as well as figuratively. For Robyn, the re-imagining of her dictation software had taken place without her intervention, indeed even before she accepted the job, as the inscribing of the new programs of action had been accomplished by the Hansard organisation.

6.3   Leaping into Use

Once an actor has made the decision to use dictation software and that decision has been supported by one or more reasons to use the software, by previous experience with similar tools allowing them to imagine using dictation software in their work and by having enough IT savvy to be confident in re-imagining the software, they are ready to "leap into use".

Use of dictation software is a leap because it is not a task or technique that can be gradually introduced into an existing work process. Using dictation software requires a drastic re-thinking of an existing work process— a leap into the unknown. To make the leap easier, there are three criteria that must be met. Adequate training is needed so that a user can be guided in the re-figuring of their work and their discovery of the dictation software. Appropriate hardware and software are required. And the work being performed must be compatible with the functionality of the dictation software.

6.3.1   Training

The perceived quality of the training that the interviewees had received seemed to relate to the feelings they described towards dictation software. Most of the interviewees had a positive view of their experiences with a trainer. Olivia said that her training experience was valuable in allowing her to become proficient with the ASR software.

> "And when I first got it [i.e. dictation software] it seemed impossible so I just didn't use it for the first year and I had a lot of pain in that year and then I took six weeks off and came back determined that I was going to make it work. I had some training once a week for a while [that] was kind of organised between me and the organisation and the Comcare person. So I learned more of the commands and got it working well in e-mail and kind of okay in Word."

Yvonne had a mixed experience. When she first began to use ASR software to cope with her injury, she was somewhat at the mercy of the trainers and consultants her department had organised for her. She said:

> "The department threw a huge amount of money at it [i.e.,
> rolling out ASR to injured users]. They got a consultant in, who
> had some sort of relationship with IBM, so we trialled IBM Vi-
> aVoice for a while. But with no training and no manuals, be-
> cause he got some sort of advance copy of the latest ViaVoice
> that they hadn't produced manuals for yet. It was a ridiculous
> situation because you would say a command and you'd just
> have to guess what the commands were and you didn't know
> if it wasn't recognising it because that wasn't the right com-
> mand or was it?"

Having a manual for the speech commands is important because there
are a finite set of commands to which each different ASR application will
respond. Without a manual, Yvonne and her injured colleagues were in
the dark.

Wendy also had a poor initial experience with training but her subsequent experience was more positive.

> "I had a very inexperienced trainer, or very experienced but
> very heartless kind of girl that actually had an injury herself but
> all injuries are different and she was quite young and she just
> assumed that you give someone the knowledge, give someone
> this basic training and away they go. And I spent about six
> months in a dark corner until fortunately she went overseas
> and someone else came to train me. She was in my department
> and had an injury and had excellent skills in all sorts of areas.
> And she said well, your system is not working properly."

In Wendy's case, poor initial training left her struggling for six months
thinking the problem was with her when in fact the problem was in the
system itself.

The other interviewees did not discuss their experiences with trainers.

Training is important in allowing users of ASR software to construct useful explanations for themselves about how the software works. Because users who do not have a strong technical background are unaware of what working software is like to use, training allows them to work with an expert who can educate them. The non-deterministic nature of ASR software makes it difficult for users to understand what is going on when the software behaves in unpredictable ways. A good trainer can assist users, of varying levels of technical experience and ASR experience, in coming to terms with the vagaries of the ASR software. As Yvonne said:

> "It's very temperamental software and my experience with it
> is that there is an element of unreliability and unpredictability
> even at its best."

Being able to ask an expert, "is that normal", allows users to understand the software and what it does. By allowing users to gain a better understanding of the software, training helps the users imagine and re-imagine their use of dictation in their work. A trainer can provide targeted advice when something goes wrong, suggesting corrective action that may not occur to a user who is naive to the functionality of the software.

Training allows users to better understand the software. All software is "opaque", it is not easy to understand how it works from a user's perspective. "Transparent" objects like a bicycle, for example, are easier to understand because all the components are on display. Dictation software, like all software is a "black box". Only inputs and outputs are knowable and the mechanics of how inputs become outputs are hidden. Black boxes put users in relationships of dependency on experts (Jordan and Lynch, 1992) but training can reduce or eliminate that dependency by helping users to become experts who can see inside the black box.

6.3.2   Hardware and Software and Environment

The right computer hardware, software and work environment are essential for easy, efficient use of ASR dictation software.

The right software is dictation software that allows users to customise it, usually through macros. Customisation of the software is important for workplace use because it allows the software to be changed to create short-cuts for complex operations. For example, one of the macros used by the Hansard editors allows them to perform the complex formatting required in the Hansard document by speaking a very short phrase, "go speech seat" where seat is the particular member's seat whose speech they are transcribing. The injured interviewees used macros in their work too, particularly for interfacing between the dictation software and other software that they used in their work.

The minimum computer hardware requirements for dictation software, like all software, are printed on the shrink-wrapped box the software is sold in. The right hardware is far above the minimum requirements for use of dictation software. The minimum hardware is adequate for using dictation software in a stand-alone situation where no other software is being used simultaneously. In every case, the interviewees were running more than just the dictation software for their work. With more software applications running simultaneously the load on the system is higher and the dictation software cannot gain access to the hardware resources it requires to run correctly. When dictation software cannot gain access to a sufficient amount of resources, recognition accuracy drops to unacceptable levels. The interviewees who were using dictation software in their daily work had high-end computers with very fast processors and more than double the typical amounts of memory.

However, efficient ASR use requires more than a fast processor and a lot of memory. The sound-card in a computer and the microphone used to pick up the user's speech play a large part in the ease-of-use of dictation software. Most, if not all, dictation software comes boxed with a microphone that is, at best, only adequate. The interviewees all used expensive after market headset microphones in their work because they said that the after market headset microphones have superior noise-canceling capabilities compared to the boxed microphones and also provide clearer transmission of the analog voice signal to the computer.

The sound-card in the computer is also important. The analog signal from the microphone must be encoded into a digital signal before it can be processed by the ASR engine in the dictation software. The quality of the sound-card influences the accuracy of the analog-to-digital encoding. All computers produce a lot of electrical noise which can interfere with the encoding, degrading the quality of the digital signal when the sound card is insufficiently shielded. The quality of the digital signal has a great influence on the recognition accuracy that the dictation software can achieve. Some microphones come with built-in analog-to-digital encoders which allow them to bypass the sound card and any shielding problems that may be present.

The interviewees all credited their high-end hardware with allowing them to use dictation software productively. They also said, with the exception of Robyn and Margaret, that acquiring the high-end hardware and software was a task that involved much negotiation with their superiors. The injured interviewees were typically not senior in their organisations and so were not automatically entitled to expensive computers.

The environment in which the user works also influences their success in using dictation software. Even with after market noise canceling microphones, external noises can adversely affect the recognition accuracy of the software so working in as quiet an environment as possible is essential. Of the interviewees, only Margaret was senior enough to have her own office. The other interviewees worked in cubicle spaces or shared offices. The shared offices at Hansard were specially configured to allow near-ideal dictation conditions. Each editor faced into a corner which had special sound-proof panels installed to minimise echos (see figure 4.1 on p62).

Some of the injured users had taken steps to create better environments for their use of ASR. Olivia sat apart from her colleagues and had erected soft material-covered partitions to further minimise noise. Wendy, who was not working at the time of the interview, said that her need for an appropriate environment in which to use dictation software was a cause of tension with her colleagues. The other injured users reported that their colleagues were supportive of their need for a modified work environment.

6.3.3 Fit with Work

The injured interviewees all reported that they had some difficulty in using dictation software in their work. The problems were directly related to dictation software: imperfect recognition, difficult integration with other software, difficult integration with aspects of their jobs. The difficulties reveal that the ASR applications they are using are not a direct replacement for the use of keyboard and mouse. Though ASR dictation software is sold with the promise of being a replacement for typing, actually using a dictation software package is not as simple as just talking. Because the injured interviewees are compelled to use the dictation software, they must reconfigure the software and inscribe new programs of action on it, and themselves, in order to work productively. The Hansard users, in contrast, are in the advantageous position where they do not have to imagine and inscribe new programs of action on the dictation software because the Hansard *organisation* has performed the translation.

The injured interviewees had different levels of dependency on the ASR software. Some were not able to function without it:

> "I can't work without the software. I'm not completely dependent on it but I can only go for a few days before problems kick in. If one of those days is a very busy day if I had a lot of submissions to do or deadlines, deadlines are a killer for overuse injuries. Then I think one day would be it. One day can exacerbate a workplace injury and [then you're] pretty much back onto anti-inflammatories and physio time."

Other injured interviewees were able to free themselves from using the software all the time.

> "I used to use the program to do everything, moving the mouse around, every possible command. Now I use the mouse and I type [...] key shortcuts. That's one of the nice things about it: you can use it or not use it as much as you need to."

The Hansard users were only tied to using ASR software for turning the parliamentarians' speech into text and were able to work with their computers using keyboard and mouse for the rest of their work.

What these examples show is that dictation software, as currently implemented, is quite flexible in *when* it is used. In no way does the implementation force a user into any more use than is required by their need, whether that need is pragmatic, as in the Hansard case, or caused by injury.

Deciding when to use the software is influenced by the software allowing a user to actually do the work they need to do. For example, a graphic designer could not perform the bulk of their work with ASR because graphic design work relies on mouse (or drawing-tablet) use which cannot be replaced by dictation software. While none of the interviewees were in the position where their work was impossible to achieve with ASR, they had varying experiences with how closely their work and the functionality of the dictation software were aligned.

Olivia said her initial experience of using the ASR software was in a job where it didn't really fit with the way she had to work.

> "I learned more of the commands and got it working well in
> e-mail and kind of okay in Word, but there I was mainly doing
> statistical analysis and version 5 didn't work very well. Most
> of our Word documents are [really] long so it just meant that
> there was too much of the work that I just couldn't do."

Long documents are difficult to work in with dictation software because the software's mechanisms for navigating long documents are poor. Similarly, the cursor-positioning commands are inefficient. Wendy also found using dictation software to work with long documents particularly difficult.

Yvonne originally worked in a job that was based around working with numbers but moved jobs to better fit with what she could achieve with ASR.

> "So, why did I start using it? I got quite a serious overuse in-
> jury and my work used to involve writing about statistics [...]

and that was very keyboard intensive work and very mouse intensive too and I actually left that work to move to a policy area because it's less keyboard intensive work and the work is mostly words and not the mouse. I moved over from [one department] to [another department] because it was better for my injury."

It is not the case that the current ASR software programs cannot work with numbers, spreadsheets in particular, or long documents. That two of the interviewees specifically moved jobs to better use the software in their work indicates that the way the ASR software works with numbers in spreadsheets and long documents is not as efficient as using a mouse or keyboard. Two reasons that ASR may not be appropriate for interacting with a spreadsheet are that (1) navigating a spreadsheet is a two dimensional activity, and navigation is not a task that ASR has been designed for, and (2) using a spreadsheet often involves creating "formulas" which are combinations of keywords, abbreviations and symbols, strings which are difficult to express efficiently in speech to a computer.

Other interviewees were fortunate to have jobs that aligned closely with what the ASR software allowed them to achieve.

What these experiences in fitting the ASR software to work show is that the software is not able to meet everyone's needs and that there are times when the work that the user-and-software pair is asked to perform is beyond the capability of the pair and this is usually due to the somewhat impoverished facility of moving a cursor by voice. The interviewees who switched to new jobs left work where they spent a lot of time navigating documents (text or spreadsheets) to take up jobs where they were better able to use the capabilities of the software.

Having work that fits with the capabilities of dictation software is a critical juncture on the trajectory of using dictation software in the workplace. If there is no fit between the software and the user's work, nothing else will overcome that problem. The necessity of fit with work influences the actor-network in which users of dictation software find themselves by forcing the enrollment and translation of many elements if the dictation software actor-network is to be sustained.

6.4   Sustaining Use

Having imagined using ASR dictation software in their work and leapt into using it, users must continually sustain their use of dictation software by balancing many competing factors. They must stay IT savvy, their reasons for using the software must remain, they must retrain, they must maintain their hardware and software in working order and their work must continue to fit with the capabilities of the software. Every day, users of dictation software undergo a trial of strength in which the actor-network they have arranged around themselves is tested. Continued productive use of dictation software depends on being able to maintain the actor-network against forces that would destabilise it.

6.4.1   Necessity of Use

Because it can be so difficult to use ASR dictation software in the workplace, an important aspect in inscribing users is for it to be necessary to use the software. Not all the interviewees could have done their work without using dictation software. For the injured users the necessity came from their desire to avoid inflaming their injury. For the Hansard users the necessity came from needing to turn parliamentarians' speech into text and being able to find very few people with the necessary typing speed and accuracy to do the work by hand.

Using dictation software every day shows people, who may be skeptical about the use of dictation software, that it works. By making it an indispensable part of their daily work to use dictation software the users ensure that their work and their use of the software become their way of working. Only having one way of working means that the users are compelled to maintain all of the connections and hard-won concessions that allow them to work.

Being injured was the reason that six of the interviewees needed to use dictation software. *Remaining* injured and wanting to avoid aggravating their injury is something that allows them to sustain their use. People who are not as badly affected by workplace overuse injuries as the interviewees often do not sustain their use of dictation software because they can work around their injury. As Margaret said:

> "There are disciplines that you've got to follow when you're using this. There are inconveniences, there are logistical issues and unless you have the impetus there, you are not going to use it."

The issue that put the injured interviewees on the trajectory of making dictation software useful in the workplace is also the impetus that keeps them there.

For the Hansard users, the impetus that makes them keep using dictation software is the need to turn parliamentarians' speech into text which can then be edited and formatted into the Hansard document. Dictation software has been chosen by the Hansard organisation as the one way for that transformation to take place. There are a small number of Hansard editors who do not use dictation software in their work, preferring to use Computer Assisted Transcription or another method, though most editors currently use dictation software.

6.4.2 Technical Support

Translating the interests of support providers, be they IT support or Occupational Health and Safety (OH&S) to align with those of injured users is very difficult. Users of ASR dictation software require technical support that understands the characteristics of dictation software. Good technical support is vital for even the most technically experienced user as most, if not all, organisations do not give workers full control over the configuration of their computers. In many cases where there is a problem with a system, a user will require someone from the IT department to investigate the problem and effect a solution. Technical support providers prefer to look after systems with a great deal in common because it makes their work easier. Supporting a few users, or even a single user, out of hundreds of users, is a low priority.

Even though many of the interviewees possessed above average technical ability, they all spoke of the need for better technical support.

Olivia said:

> "They have delivered the hardware and installed [the software] originally but they took their time about it."

Olivia's IT support was out-sourced. Typically, out-sourced IT support is unwilling to agree to support dictation software users without additional support fees from the agency where the user works. Yvonne, who worked as an ASR trainer, said that users who worked in agencies with out-sourced IT support generally found it harder to integrate the dictation software with their work.

Margaret was very positive about her department's IT support and I asked her if it was internally provided. She said:

> "We do, we do. Thank God. They decided earlier this year that they were going to go to an out-sourced help-desk. Most of them are still contractors but they are based in the building which makes a big difference. I've been through more IT officers supporting voice software than I care to remember and the quality has been variable and the support that they are given to prioritise this kind of work is very variable as well, and that's a difficulty."

I asked Margaret for an example of the service that she had received. She told me how she had returned from holidays to find her ASR software had stopped working properly. She had approached her IT support providers but they were evasive.

> "I tried to get it fixed. For the first week they kept saying 'we're too busy with other things on'. They make a time to come and then they wouldn't make it. You would ring them up and they would say, 'Oh! Something came up'. After two weeks my injury had exacerbated and I was back seeing the physio twice a week. They did fix it after the injury had been exacerbated, but we had to go through another ComCare[3] process. It was just

---

[3]Comcare is responsible for workplace safety, rehabilitation and compensation in the Commonwealth jurisdiction (ComCare Website, 2005)

> a huge thing. Since then they've been a bit better, but you go
> through periods where service was quite good and then there
> will be a loss of corporate memory as to what happens when
> you don't provide service and it does take a re-injury."

Margaret told me that she was trying to have some succession planning
implemented because the current technician in charge of ASR support was
going to leave soon and she did not want all of the knowledge that he had
built up to leave with him.

At the time of the interview with Margaret she had recently given her
new computer over to her IT support team because a new Electronic Document Management System (EDMS) had been installed for the whole department and it did not work with the dictation software that Margaret
used. She was annoyed at this because she was "effectively disenfran-
chised from the department document creation, sharing and storage system". She told me that in the planning stages of the introduction of the
EDMS:

> "When they were first talking about purchasing this EDMS I
> went to the people and said will it work with voice-activated[4]
> and they said, 'yeah, it'll be fine, we know all about you'. When
> it comes to the crunch, 'oh, why didn't you tell us?'. Well I did.
> Two years ago. And I've got the files to prove it."

The poor experience of the injured users with IT support is in contrast
to the experience of the Hansard users. The Hansard department has its
own internal IT support team who are dedicated to supporting all of the
systems that the Hansard editors use. In addition to dictation software
there are several other software packages that are essential to the production of the Hansard document. If any system fails at Hansard, the day's
work cannot get done so good IT support is crucial to the goals of the organisation. By tasking IT support with supporting the goals of the organisation, the Hansard department has enrolled them in the actor-network

---

[4]Which is what Margaret called dictation software.

and re-aligned the goals of the IT support people with that of the organisation.

Because the injured users are somewhat outside of the norm in their organisations, their IT support people are not enrolled in the specific dictation software actor-network that the injured users need them to be. The injured users must convince their IT support person (or people) to provide support for dictation software. Supporting dictation software is typically outside the knowledge base of the average IT support person so performing the "interessment" requires that the dictation software user convince the technician to learn a new software system and then provide technical support. If the technician works for an out-sourced IT provider they will have a contract that specifically states what they can and cannot provide support for and as it is highly unlikely that dictation software will appear in the contract the dictation software user may be unable to get technical support. In Olivia's case because she was one user in one hundred she was a low priority even though she was covered under the contract.

Yvonne told me that, in her work as a trainer, she had noticed that there was poor communication between IT and OH&S regarding support of injured users. She thought this was partly because people who were attracted to working in the OH&S field were not technically minded and partly because "IT people have no stake in Occupational Health and Safety outcomes so this is just extra work for them. It's not part of their brief. No one is going to tell them, 'you've done a great job this year because you helped three people remain productive' and often there is no commitment at a high level to make those two areas talk."

Yvonne was aware of the importance of receiving good support from OH&S and IT. She told me the story of how she had worked at a department where it was difficult to get any assistance and how the section of the department that she had worked in had been taken over by another department that had a team of people dedicated to supporting injured and disabled users.

> "She was a professional occupational therapist so she had professional training which is unusual in Occupational Health and Safety areas, so they just did their job amazingly well. They

> came in and they met with us all and they had a whiteboard and they wrote up what they had to do to get us all up and running and it all happened and they seemed to have funds. She came over one day and handed me a graphics tablet which at that time cost about $700 and said, 'here, try this, this might be useful'. It was just amazing. And we were so grateful after seven months of [her previous department] trying to fix it."

This experience of good support occurred early in Yvonne's use of ASR software and seemed to have influenced all of her future dealings with support people from the IT and OH&S areas.

More than the other interviewees, Yvonne had the view that good technical support was a right, not a privilege. She would act as an advocate for her trainees to get them better support.

Generally, the injured interviewees saw that despite frequent poor support, occasional good support meant that good support was possible. It was that view that led them to agitate for better support, in the case of Yvonne's lobby group, or for better continuity of support, in the case of Margaret's desire for better knowledge transfer when a support technician moved on. Because of the frequent lack of support, most of the interviewees had become more self-sufficient regarding their use of ASR software. It was clear that Yvonne, in her work as a trainer, encouraged this and helped build it into the views of her trainees while simultaneously encouraging them to try to get better technical support.

6.4.3 Becoming the Conductor

In order to truly sustain the use of dictation software in the workplace, someone must become a "conductor", controlling all the heterogeneous elements that must be brought together in order for dictation software to be useful. An actor must take on the enrolling, translating and inscribing of other actors. The one actor sets the script, the program of action, for the other actors to follow.

The injured users interviewed for this research are the conductor but they are typically not powerful enough to convince other actors to assist

them. They must enroll other more powerful actors in their network and
have the more powerful actors perform the work that makes dictation software useful. Several of the interviewees had found a powerful ally in the
form of a collective of other injured users within their agency.

Olivia, who had spoken of her less-than-ideal technical support later
told me that her department had a "special-needs group" who acted as
advocates for injured or disabled employees.

> "We have a special-needs group and they are incredibly sup-
> portive. [A technician attached to the special-needs group] will
> come around and if Word isn't working with Dragon he will
> fix it or he will sit next to you and see which commands are
> not working so in terms of that it is brilliant. They will take up
> issues for you too, for example I had an issue with training…"

Olivia described the training courses that her department put on for
employees. They were delivered in a dedicated training room and were
not appropriate for an ASR user due to the environment—Olivia would
have to use her ASR software in the room with other non-injured colleagues which would be disruptive for the others as Olivia would be talking and it would be difficult for Olivia because it would be a less than
ideal environment. The trainer has suggested that he could come to Olivia
and provide her with one-on-one training. Olivia's department has made
it possible for her to have a voice with their out-sourced IT provider by organising the special-needs group. Without the special-needs group, Olivia
would be a single user asking for special assistance, which is unlikely to
be successful in an out-sourcing situation.

Margaret's department also had a special-needs group, except within
her department it was called the "Adaptive Technology Workgroup". At
the time of the interview it had only recently been established and was
intended to "act as a bit of an intermediary between users of adaptive
software and IT and senior management". In Margaret's department there
were employees who used software other than ASR to allow them to function productively within the organisation despite an injury or disability.
Margaret said that there were employees who were blind as well as sev-

eral other ASR users.

Margaret was, more than any of the other interviewees, a high-profile person in her department. She was highly educated and very experienced in a very specific field and she had similar difficulties to the less high-profile interviewees. As a user of a minority software package that is difficult to use and can often be cantankerous she was at a similar disadvantage to the other ASR users in this study. She was a productive user of the software, despite her difficulties.

Yvonne was one of the most pro-active interviewees. She told me that after a number of IT staff and a "disability access coordinator" left, leading to a decrease in the quality of support, she and a number of other ASR users formed a lobby group.

> "It made me realise that it was just those two people doing a very good job in IT and the disability access coordinator. And also in that agency there were quite a lot of people using Dragon; about 30 people had installed it but there were probably really only 10 to 15 core users because a lot of people have a go and then don't continue. And we formed a lobby group too and we got a lot of things and we achieved an enormous amount."

I asked Yvonne for an example of what the lobby group achieved with regard to increased support.

> "At that time my system was working well and they had people on all different versions. Some people were on version 2, others were on version 3, 4, 5. If you were a new user you would be put on the latest version but old users they would just leave them on the older version. We said, 'can you put all Dragon users on Windows 98 because it doesn't work on this version of NT' and they said, 'no we can't do that, we couldn't possibly do that'. Even though we'd won the diversity award. And we said, 'well, what will your defence be if somebody takes legal action?' And then suddenly, and it's hard to say it wasn't that, that threat of legal action, suddenly there was

this huge lot of money, everyone got rolled out onto version 5 and new hardware and they bought VoicePerfect [an ASR software vendor and training company] in with multiuser licences and training for everybody and that must have cost a fortune. And I think that was the lobbying."

Yvonne has since left the Public Service and works as an ASR trainer. In her role as a trainer Yvonne said she often acted as an advocate for injured users. Yvonne would typically be brought in by the department that employed an injured user and as an outside consultant she had a lot of power to effect change. Yvonne willingly joined the actor-networks of her trainees and helped them conduct the elements they required in order to be productive.

All of the injured interviewees said that despite their various lobby groups persuading IT or OH&S representatives to effect change they still found it difficult to get timely assistance. Yvonne told me that the "best place [for good service] is small high profile places with lots of money and where the clients are high profile. So you're obviously less likely to mess around with a senior legal officer than with an ordinary AS05 [a middle level public servant]". In other words, lobby groups of low-level employees are still less important than a few high-level senior people.

The interviewees all had mixed experiences in conducting the necessary elements except for Robyn from Hansard. At Hansard, ASR had been adopted as part of the work practice and had full support and commitment from the higher management. It was apparent that the support from management had filtered down to the IT support because Robyn never raised the issue of support, from any perspective, during the interview. Instead, IT was directly involved in the ASR actor-network at Hansard, developing macros and otherwise supporting the users on an ongoing basis.

6.5   Summary of Analysis

Using dictation software productively in the workplace is not easy. Giving a person a shrink-wrapped box of software does not allow them to integrate that software into their work. It emerged from the interviews that productive use of dictation software is the end-point of a trajectory (Fitzpatrick, 2003; Graham et al., 2005) that must go through various waypoints before it reaches its end.

When a person decides to use dictation software in their work they first need a compelling reason to even consider it. In the case studies presented here the reasons are, in the case of the injured users, injury that prevents conventional keyboard-and-mouse computer use, and in the case of the Hansard users, because dictation software allows a work process that would not otherwise be possible. Previous experience with similar tools can be of assistance when a user is considering using dictation software. In the Hansard case it was not a single user who considered the use of dictation software but the organisation which saw the dictation software as sufficiently similar to older ways of working that it could replace them. As the cases of the Hansard department and the injured users show, the will to consider the use of dictation software does not necessarily rest with a human but can lie within a non-human actor. Finally, to consider using dictation software an actor must be sufficiently aware of the possibilities afforded by technology—they must be IT savvy.

Having decided to use dictation software a leap into use must take place. As with deciding to use dictation software, the elements that allow such a leap to take place are heterogeneous. Training must take place in order to show the individual user(s) of the software what is possible and what cannot be done. Training augments a user's self-discovery of the software. The correct hardware and software must be provided, allowing the dictation software to work at its best. An environment that allows for the use of dictation software is also necessary. Most importantly of all, the work that is being performed must fit with the capabilities of the dictation software in order for a leap into use to take place.

After deciding to use dictation software and leaping into using it, an actor is at a critical juncture. They have enrolled diverse heterogeneous actors in an actor-network dedicated to a single person using a particular piece of software. Each heterogeneous actor has been assigned a script to follow. The hardware and software must work as promised, the user's work must fit with what the software can achieve, the training must have given the user the necessary skills and knowledge to use the software. If any actor does not follow their script, the actor-network will collapse and the user will be unable to maintain their use of dictation software. The actor who first imagined the use of dictation software must sustain the actor-network.

The injured users' actor-networks are not very stable (see Jordan and Lynch, 1992, for another example of unstable actor-networks). They are continually negotiating with the heterogeneous actors in their networks in order to sustain them. The actor-network at Hansard is much more stable. The actor who has the most power in the network, the Hansard department, is much better able to conduct the network than the single injured users. Even the injured users' lobby groups do not have the power of the Hansard department to effect stability.

Through the interviews it emerged that there were several elements that were particularly helpful in sustaining the use of dictation software. Use of dictation software must be necessary—the user must be compelled to use it—there can be no other option. Necessity of use makes it in the user's/actor's interest to maintain the network. Technical support is also vital in sustaining the use of dictation software, both from IT and OH&S. IT support has the ability to change things that users typically cannot but that may be necessary to change for dictation software to work. OH&S support can act to influence policy within an organisation by being dedicated to the needs of employees who are disabled.

Finally, a user/actor must take charge of maintaining the relationships between the heterogeneous elements necessary for the sustained use of dictation software. Conducting the elements in this way requires power that is often beyond the reach of single users. Getting the necessary power can take the form of collective action and finding strength in numbers. Injured users who have grouped together have achieved more than what

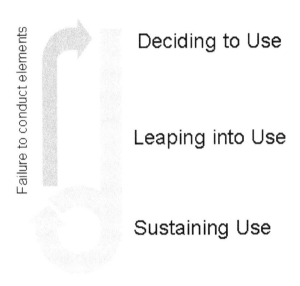

Figure 6.1: A model of the trajectory of using ASR in an organisation.

was possible alone. The Hansard department is a good example of an actor sustaining the use of dictation software. The Hansard department is powerful enough that it can compel the heterogeneous actors to follow the scripts assigned to them.

Using ASR in an organisation is a struggle. It is a struggle to constantly maintain the relationships between the heterogeneous elements necessary to enable the use of the software. If, at any stage of the trajectory, a destabilising element is introduced it is possible that users will find themselves sent back to the beginning of the process (see figure 6.1) unless the user has enough power to enroll the destabilising element into their, that is the user's, actor-network.

The next chapter returns to the ACT Magistrates Court and considers the future use of ASR at the Court. The trajectory of use of ASR developed in this chapter is used in a proposed design for a future ASR application at the Court.

CHAPTER 7

Analysis of the Court situation and Design of a new sentencing system

In this chapter I move from an ethnography of automatic speech recognition (ASR) users (see chapters 4 and 6) and an ethnography of the ACT Magistrates Court (the Court) itself (see chapter 5) to a design, or sketches of a design for an ASR system at the Court. Moving from ethnographic studies of different technologies and workplaces to design has been the subject of much research (see chapter 3). This chapter moves from an ethnography of the Court to a design for an ASR system for the Court.

This project began when the Chief Magistrate of the Court asked if an ASR application could be developed for use in the Court during sentencing (Kraal et al., 2004). I initially approached the project using a software design paradigm however it soon became apparent that designing an ASR application for the Court required considerably more knowledge about the Court and about ASR than I possessed.

The literature on the design of ASR systems (Weinschenk and Barker, 2000), while strong on the process of designing a system, did not describe how the software is actually *used* in a day-to-day environment. Where much of the ASR literature is concerned with overcoming the inherent disobedient nature of ASR software, it was difficult to find out how people actually coped with existing ASR software when they had to use it daily.

In the previous chapter I have shown how ASR is currently used in office environments and how the users of the software adapt themselves to the software and the software to themselves. I have also shown how using ASR software is not a single-user task but one that requires an extended network of support including other people, training and various hardware and software.

This chapter describes the process used to design an ASR application for the Court. The application whose design is described here is designed from observations and analysis of the work of people in the Court and from observations and analysis of how ASR users work with that software.

The specific work that is performed in the Court that the design supports is termed *communicating outcomes*. Communicating outcomes (see section 5.1) is a process that is distributed in space and time throughout the Court. It begins when a magistrate makes a decision on an (interim or final) outcome which they speak aloud to the Court and then record by hand on a *bench sheet* which is inserted into a *defendant's folder* and which then travels to various locations throughout the Court building depending on the decision. If the outcome is an interim one, various paperwork is produced from the outcome recorded on the bench sheet which ensures that all interested parties are informed about the next step in the process. A sentence, determined by a magistrate, is the final outcome in any particular case. A sentence outcome, recorded on a bench sheet and placed into a defendant's folder works its way through the Court, passing through various hands before stimulating the production of the necessary paperwork to enact the sentence.

Recording outcomes by hand while court is in session is a time consuming process as many outcomes are long and detailed. The reason for investigating ASR for the Court was to allow the magistrate to avoid recording outcomes by hand and instead to speak decisions on outcomes.

In the next section (7.1), several scenarios are presented that describe various visions of ASR at the Court. The scenarios are used to illustrate the difficulties in using ASR at the court (section 7.2) and stimulate re-imagining of ASR for the court (section 7.3).

7.1 Scenarios for Automatic Speech Recognition in the Courtroom

In this section I will first explore what the Chief Magistrate wanted from an ASR system in the ACT Magistrates Court. In the second part of this section I will examine how my research into ASR and my investigations into the Court's work process combine to make the Chief Magistrate's envisioned system impractical.

Later in this chapter I will describe the areas of the Court where my research shows ASR could be successfully used and why I believe those areas are suitable for ASR.

The next section uses scenarios (Carroll, 1995) to describe the Chief Magistrate's vision for ASR at the Court. Two further scenarios are presented. Both scenarios are caricatures: they represent extreme cases for the technology in the Court. The use of scenarios in this way (Bødker, 2000) is to stimulate ideas and make clear the potentials and problems of ASR in the Court. The first scenario represents the best case where all the technology works smoothly and there is no disruption to the work process. The second scenario is a worst-case scenario where there are a lot of problems. By contrasting the two scenarios I show where the potential fault-lines are for implementing ASR at the Court.

7.1.1 Background to the Scenarios

We were approached by the Chief Magistrate of the ACT Magistrates Court to investigate the introduction of ASR technology to the courtroom for use by the magistrate in the process of communicating outcomes (see section 5.1 for more on outcomes). The Chief Magistrate asked for an ASR system that could replace his existing manual system of handwriting and rubber stamps. When the time comes to pronounce sentence, a magistrate has the option of using a one or a combination of large rubber stamps (see figures 5.4, 5.5 and 5.6) and handwriting to record a sentence. They will also speak the sentence aloud. The Chief Magistrate thought that, since he was speaking the sentence, an ASR system could be employed to record what he had said and remove the need for him to record sentences on paper. His main reason for wanting an ASR system was so that he could save time. Writing outcomes down is time consuming, particularly as one defendant may be appearing on many charges, each of which will require a decision from the magistrate. A magistrate will often decide to waive many of the individual charges and sentence a defendant on a small selection of the total number. The waived charges still require a stamp and some writing and so still take up some of the magistrate's time that could otherwise be used to hear cases.

The process of communicating outcomes is a highly charged moment in the Court when the magistrate speaks an outcome for the case that he or she is hearing. Each case may have more than one outcome. An outcome may be a sentence, for example a fine or jail term, or it may be the decision to set a case over to allow all the parties to the case more time to gather relevant information. An outcome may also be a procedural decision specific to the Court such as a request by the magistrate for any number of specialised reports that are used to inform the actual sentence when it is finally delivered. The magistrate's speech act changes the world. It determines whether a defendant can leave the courtroom or is returned to the cells.

After some preliminary ethnographic work at the Court it emerged that the magistrate's act of speaking an outcome was not an event that was self contained but was the beginning of a process distributed in space and time throughout the Court[1] and led to the recording of an outcome in many different places and for many different purposes. This contrasted with the Chief Magistrate's view of the process as one which was enacted by him and contained within the courtroom.

In the next sections, I describe the "fantasy" scenario that prompted the Chief Magistrate to contact the University of Canberra regarding ASR. While there was never a time when he explicitly said to me "this is my dream for speech recognition" it became apparent to me that the following scenario is very much what he had in mind. Following the fantasy scenario is a worst-case scenario that shows how the same basic imple-

---

[1]See chapter 5 for a description of the work process of the Court.

mentation could be disruptive to the Court. These scenarios are not in-
dicative of my design for ASR for the Court. Instead they show a positive
and negative view of ASR from the magistrate's perspective to illustrate
the demands of a future application (Bødker, 2000). These scenarios are
not drawn from real examples but are constructed. The scenarios act as
"means to hold on to situations and how they may be changed because
of a design" (Bødker, 2000). In both the positive and negative case, the
scenarios are extreme, very good and very bad, to show the "full-blown
consequences" of an ASR system.

7.1.2 Scenario: Techno-Utopia

It's 9.30am on a Tuesday as Rob, Chief Magistrate of the ACT, enters the
courtroom. He sits down at the bench and court begins. On the bench are
several objects: Rob's favourite coffee-mug, a carafe of water and a glass,
a few pens, an array of tiny microphones embedded into the small shelf
above the surface of the bench and a touch-screen that's about as big as
a hand-held computer game. The microphones work together, canceling
noises from the Court and capturing Rob's speech when necessary and the
touch-screen allows Rob to trigger various modes and actions of the ASR
system.

The first few cases that appear are dealt with very perfunctorily and are
all set over to another date. Rob does this in concert with the List Clerk
who advises him when the next available dates are for the particular sort
of cases that appear. Rob's Associate, Claire, organises the cases in this
way as it suits Rob's way of working. Once Rob and the List Clerk have
found a suitable date, Rob uses the touch-screen to trigger a *recognition
event* that allows him to speak the date for the next part of the case to the
Court. Speaking the outcome records it.

The next cases involve people who have been in the lock-up overnight.
Rob usually makes a judgment on these cases, often just a bail arrangement but if someone pleads guilty he will sentence them on their first appearance if the sentence is simple and not severe.

The first difficult appearance today is a Mr Tailor who was in a street brawl last night and has been in the lock-up since about 2am. The public prosecutor hands Claire a police report on the incident that Claire hands to Rob for him to read. Mr Tailor's lawyer says that the fight was uncharacteristic and that Mr Tailor is a member of society in good standing who has been employed as a carpenter since he left school at 16. Rob says that the report indicates that Mr Tailor hit three people, including a woman, and that he swore at a police officer. Rob says that these are fairly serious charges and that he will have to sentence Mr Tailor.

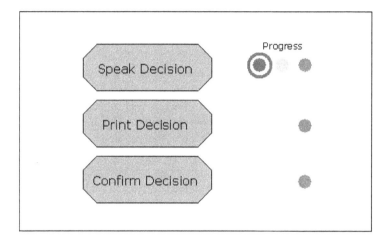

Figure 7.1: The touch-screen interface in the ready state.

Mr Tailor's lawyer and Rob have an exchange that results in Rob postponing sentencing to a date in three week's time. To make this decision official, Rob touches a button on a small touch-screen (see figure 7.1) mounted on the bench. The button is labeled *speak decision*. The button changes colour from grey to green, showing Rob that the system is ready. Rob says, "Decision in case 54897," and then says the words of the bail agreement, "the defendant is released on bail, recognisance self in the amount of $1000 to reappear three weeks hence"[2]. An indicator next to the

---

[2]These are not the exact words used in a real bail agreement.

button turns yellow and then green, indicating that the decision has been recognised. Rob taps another button labeled *print decision*. A small laser printer in the bench produces a piece of paper with the decision printed on it. Rob checks that he is happy with the wording, signs it and places it in the bench sheet folder. He taps the next button in the touch-screen, labeled, *confirm decision*. Next to Claire, a laser printer comes to life and produces three identical pages. Claire hands one to each lawyer and one to Mr Tailor. These pages contain the text of the decision and the date of Mr Tailor's next court date. Pressing the confirm decision button has also added the decision to the Court's computer system. The touch screen goes back to its initial state, ready for the next case, as Claire calls for the next defendant.

7.1.3 Scenario: Dysfunctional Dystopia

It's 9.32am on a Monday as Rob, Chief Magistrate of the ACT, enters the courtroom. He sits down at the bench and court begins. As Claire, Rob's Associate is calling the first case, Rob plugs himself in to the speech recognition system. A lapel microphone is sewn into the black gown that Rob wears and it needs to be connected to the system.

The first case today is a Mr Jones who caused a car accident last night while he was drunk and has been in the lock-up since about 2am. Mr Jones is pleading guilty on all charges. The public prosecutor hands Claire a police report on the incident that Claire hands to Rob for him to read. Mr Jones's lawyer says that the drunkenness and accident were uncharacteristic and that Mr Jones is normally home looking after his four children by 9pm. Last night Mr Jones had attended a party at a local club and made a mistake in driving home intoxicated. Rob says that the report indicates that Mr Jones hit two cars and resisted arrest and that these are fairly serious charges, so he will have to sentence Mr Jones.

The defence counsel assents to Rob passing sentence immediately. To make the sentence official, Rob touches a button an a small touch-screen mounted on the bench (see figure 7.1). The button is labelled *speak decision*. Nothing happens. Rob taps the touch-screen again and this time it changes colour from grey to green, indicating that the system is ready. Rob says, "Sentence in case 86572," and then says the words of the sentence, "the defendant is found guilty on all charges and is sentenced to three months imprisonment to be suspended forthwith and is released on a good behaviour bond of $1000"[3]. An indicator next to the button turns yellow... and stays yellow, indicating that the decision parser has not been able to correctly determine the sentence. This usually means that the recognition engine has misrecognised a word so that the spoken sentence is not in a form that makes legal sense. Rob hates repeating sentences when the system gets them wrong because he thinks it makes him look foolish which is not a good way for a magistrate to look. Rob taps the yellow speak decision button again and repeats the sentence. Just as he's finishing, someone in court sneezes! At least half the time, a sneeze or cough from the gallery will ruin the speech recognition of the decision. This time, though, the button turns green so Rob taps the *print decision* button. A small laser printer in the bench produces a piece of paper with the decision printed on it. Rob checks the wording, but the system has misrecognised the length of the sentence and the amount of the bond. Why the decision parser can't check these things, Rob doesn't know. He supposes that different amounts are equally legal, even if they are wrong in this instance. It's often the case that when Rob gets a yellow from the speak decision button that the system has also got something else wrong. Rob slides his chair closer to Claire's desk to ask her to try to fix what's gone wrong but he feels the microphone cord tension as he reaches its full length, still not quite close enough to have a quiet word with Claire. So instead he glances down at Claire and lifts his eyebrows significantly. Claire taps a few keys, giving her access to the transcript of what Rob's just said, and begins editing the transcript. The system allows Claire to edit the transcript of the spoken sentence only when it's been parsed correctly. When Claire's done she nods at Rob and he taps the print decision button again. The decision comes out of the printer and Rob signs it and places it in the bench sheet folder. He taps the next button in the touch-screen, labeled, *confirm decision*. Next to Claire, a laser printer comes to life and... nothing.

---

[3]These are not the exact words used in a real sentence of this type.

Claire leans over it and sighs. Paper jam. She flips covers and latches and pulls out a mangled piece of paper. She gives Rob a small nod again and he taps the confirm decision button. This time the printer produces three identical pages. Identically faulty. The toner cartridge in the laser-printer has run out.

Claire whispers to Rob that they have a problem and Rob says to the court at large, "let's have a ten minute recess while we get someone up here to deal with some small problems we're having". Most people in the court sigh—it's clearly going to be a long day.

7.1.4 Explaining the Scenarios

The scenarios above show how the same technology, implemented in basically the same way, can have radically different outcomes in use. In the techo-utopia scenario, everything is perfect, the interaction is virtually seamlessly integrated into the business of the court. In the dystopic scenario everything breaks down, including the magistrate's sense of control and prestige in the court.

Contrasting the scenarios shows that the introduction of ASR to the court does not just require a computer, but a microphone or system of microphones, a printer, a means to engage the ASR system when necessary and contingency plans when some or all of the interconnected technologies fail. Where the techo-utopia scenario shows how simple the system could be, the dystopic scenario shows that the same technologies could be tremendously disruptive not just to the large-scale running of the courtroom but also the small-scale interpersonal interactions between the magistrate and the associate, as illustrated when the too-short microphone cord prevents Rob from having a private word with Claire, reducing him to facial gestures.

Aspects of the use of ASR in the court are also problematic because of the properties of the court itself. These properties are related to the established work process of the court, the physical arrangement of the space, how the required technologies relate (or do not relate) to one another and so on.

Neither the specifically technical nor the specifically non-technical aspects of introducing an ASR system to the court are responsible for the difficulties involved in such an introduction. Solving the problems in the technical sphere but ignoring the non-technical problems does not make a future system useful or usable. Both the technical and non-technical must be considered together in order for the design of a future ASR system to take into account the complex environment of the court.

7.2 Difficulties for Automatic Speech Recognition in the Court

Using ASR productively in the ACT Magistrates Court courtroom is fraught with difficulty. The courtroom environment is complex, both from work-process and social perspectives. Automatic speech recognition technology is currently errorful in nature and its use in the courtroom will require the assemblage of a body of associated technologies in order to make it useful. In this section I use the language of Actor-Network Theory (Law, 2003; Law and Hassard, 1999; Latour, 1987; Callon, 1986b) to describe the difficulties involved in introducing ASR to the Court. I also extend Actor-Network Theory (ANT) by taking it into a speculative design space. Where it is usual to compare two existing situations with ANT, in this section I will compare the existing court situation and my envisaged future situation. To do this, I will use the ANT research frame (Callon, 1986b) (see figure 3.3).

The court and the process of communicating outcomes is a largely stable network. The proposed introduction of an ASR system will necessarily destabilise the existing network and successful use of ASR will require that a new network be established and stabilised. By problematising the existing way of communicating outcomes and proposing ASR as the solution, I am suggesting that an ASR system become the obligatory passage point (Callon, 1986b) for the system, that is, to make the ASR system indispensable.

Because I have previously argued that ASR is flawed, suggesting, as I do here, that an ASR system become indispensable to the Court does not sit easily with me. However, by viewing the introduction of ASR to the Court as a design exercise to solve the problem of introducing ASR to the Court in such a way as to make it useful, I have reconciled myself with these apparently contradictory view points.

A new network that includes ASR will require that all the actors, or elements, or components, in the existing network be re-enrolled and re-configured in the new network. Some will resist and some will comply. New actors will have to be enrolled in the new network, too. If the introduction is to be successful, the actors who initially resist will need to be convinced or compelled to join the new network.

Thinking about the introduction of ASR to the Court in Actor-Network terms allows the richness of the situation to be included in the design thinking. Using ANT sensitizes the designer to all of the actors, both human and non-human, in the network, and can reveal opportunities for the modification of technology to fit work rather than the other way around. Similarly, ANT can be used to reveal where the proposed technology cannot accommodate existing work and where design work needs to be done to reconfigure that part of the actor-network to allow for the introduction of the technology.

7.2.1 Translating Automatic Speech Recognition and the Court

Using an automatic ASR system at the ACT Magistrates Court will involve translating the ASR system and the Court. The Court's interests are the administration of the law and the accurate recording of decisions. The ASR system's "interest" is recognising speech. To allow the ASR system to pursue its interest without interference, many actors will need to be interested and enrolled in the new network. Allowing the Court to continue to pursue its interests with as little interference as possible from the ASR system means that aspects of the Court will have to change in order to preserve the main elements of the Court. How these translations will begin to be achieved is described in the next section.

7.2.2   Interesting and Enrolling Actors

Interresment and Enrollment are difficult stages to separate and tend to overlap. This section deals with both Interresment and Enrollment.

Taking the most obdurate, stable, elements of the ACT Magistrates Court first I will describe various actors and how they can and cannot be reconfigured to allow the introduction of ASR to the Court.

The most obdurate actor in the Court is the courtroom itself and, more particularly, the physical location of various participants in the court process in the courtroom (see figure 5.1). The magistrate sits at the bench which is raised above the main floor level. The magistrate's associate sits to one end of the bench and below it. The lawyers sit facing the magistrate. The defendant sits next to his or her lawyer. The witness box is toward the other end of the bench with the witness facing the lawyers. A public gallery of varying size sits behind a barrier behind the lawyers. In the Magistrates Court, because of the relatively fast pace, there are often other defendants waiting in the court. Some of these waiting defendants are in the public gallery and some are queued up in front of the barrier but still behind the lawyers having been brought up from the lock-up by bailiffs. The physical layout of the Court is something that cannot be changed to accommodate ASR.

The social world of the Court is necessarily bound up in the physical layout of the participants. The magistrate's authority is symbolised by their elevated position at the bench. The interaction of associate and magistrate is enabled and constrained by the associate's position to the side and below the magistrate. To have a discussion with the associate the magistrate must slide over to the associate's side of the bench. Being able to move along the length of the bench requires that the magistrate not be tethered by any cords to microphones that may be required for ASR.

Slightly less obdurate *in theory* is the architectural styling of the courtroom. Court Room One in the Court is of modern design with a fair amount of pale wood in place of the more traditional dark wood paneling. The ceiling is stepped and indented in various places leading to acoustic black-spots in the public gallery. The design of the Court does not seem to

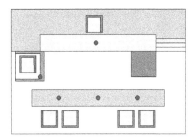

Figure 7.2: Microphone placement in the ACT Magistrates Court. Red dots represent the microphones. See also figure 5.1, 75.

affect the ability of the active players in the Court's process to hear each other. In practice, though, the architectural styling of the Court is fixed as it is expensive to make significant changes.

Various technologies are employed in the courtroom. Microphones are placed in front of the magistrate, the lawyers and the witness box (see figure 7.2), not to broadcast the speech of the players to the gallery but to allow it to be recorded. The microphones currently used are basically of "lectern" style with a small bud at the end of a semi-flexible stalk. The players in the court do not interact with the microphones due to their unobtrusive placement and because they do not hear any "foldback"[4] from the microphones which might inform them that their speech is not being fully captured. The use of microphones in the Court is therefore not a part of the work in the Court nor is it part of the embodied social world of the court, leaving the use and placement of microphones open to change and re-enrollment in a new network that will use ASR.

In communicating a decision, the magistrate's speech must be captured and the lawyers', defendant's and witnesses' speech is of no importance. This means that a method of communicating sentence need not be concerned with the microphones in front of anyone save the magistrate. By

---

[4]That is, they cannot hear what they sound like through the microphone, as one can when the amplified output of the microphone is broadcast to the room in which one is talking.

extension, the lawyers, defendant and witnesses do not need to be enrolled in a new network.

Depending on the design of the ASR system proposed for the Court, magistrates and associates may have to be enrolled in the new actor-network. From my examination of the work process of the Court, and particularly the work done during court sessions, the associate has a lot of work to do and it would not be possible to add to the workload of the associate without taking some parts away. Because much of the associate's work in court has to do with helping the magistrate manage the work process of the court, it is not desirable for the associate to perform work extraneous to that management. Similarly, the magistrate is concerned, while court is in session, with managing the process of the court and imposing new work on the magistrate, particularly when that work could involve errorful ASR, is extremely undesirable.

As has already been said, the social world of the court is embodied, at least in part, in the spatial layout of the courtroom. This is also true of the work process of the Court as a whole if that work process is seen as the administration of cases that appear before the magistrate. Defendants enter and leave with their lawyers and only those called to appear at any one time can interact with the magistrate and the public prosecutor. Other defendants and lawyers must wait in the spaces assigned to them. Similarly, witnesses may only interact with the Court, the lawyers and the magistrate when called and must otherwise wait. The spaces for waiting are important to the Court's work process as a whole but do not need to be enrolled in a ASR system for communicating decisions.

Documents of many kinds are an important part of the work of the Court. Reports from external agencies are used to inform decisions. Bench sheets with notes from a defendant's previous appearances remind magistrates about their previous decisions. In much the same way as Air-Traffic Controllers' work is embodied in paper strips (Hughes et al., 1992b, 1995) the work of the Court is embodied in the defendant's folder. As the folder moves through time in the Court, it is used at various stations to reconstruct what occurred when the defendant it describes appeared before a magistrate. Staff in the back room use the folder as part of their work in recording magistrates' decisions in the Court's computer system. They also insert documents that they generate into the folder for future reference. Magistrate's associates use the folders to prepare for court sessions and will re-order the documents in the folders to assist the magistrate to work more efficiently during court. Magistrates refer to bench sheets and reports in a particular defendant's folder during court to reacquaint themselves with their previous decisions. Because the bench sheet and folder embodies the work of the Court it is quite obdurate in the existing work process, and therefore in the existing actor-network of the court. It would be very difficult to replace it or significantly change it to accommodate a ASR system.

A tension exists within the use of the defendant's folder. The Chief Magistrate wanted to do away with the time-consuming act of writing down decisions but, as I have said above, writing down decisions is important for the future reference of the magistrate. Finding a solution to this tension is the main task of designing a ASR system for the Court. Doing away with the act of writing down decisions on outcomes may actually make the work of a future magistrate harder as there will be no record of the past.

The diverse actors who make up the parts of the Court's work process for communicating decisions that take place outside the courtroom will also need to be interested and enrolled in the new ASR actor-network. From a system design point-of-view, the easier it is to enroll the "back room" actors, the easier it will be to introduce the new system. The human actors in the back room are the monitor, the after-court person and the list clerk, all of whom have an interest in what the magistrate says and how decisions are communicated.

The monitor uses a computer system to annotate and mark-up the recording of what is said by everyone in court. Making the monitor part of the new design would be desirable because they already deal with the magistrate's speech. Enrolling the monitor's computer system and the audio feed it relies on may also be necessary.

The after-court person is someone who relies almost totally on the defendant's folder to perform their job. Any change to the folder changes the work of the after-court person because of their reliance on being able to reconstruct the decisions made in court. The work of the after-court person depends on the work of the Court being embodied in the defendant's folder.

Enrolling the after-court person in the ASR actor-network will require that the new actor-network benefits them. As the current after court person said, the work is "bulk work", that is, there is a large volume of work to do, yet the work is skilled as it requires interpretation of magistrates' decisions for translation into codes to enter into the Court's case-management system. Getting the after-court person interested in a new system will not be easy and convincing (enrolling) them that the new system will work and be beneficial will be similarly difficult.

The list clerk relies on what is said by the magistrate in court but is usually in court to hear what the magistrate says. Because an ASR system depends on the magistrate's speech, the list clerk should not have to be enrolled in an ASR actor-network.

Enrolling ASR technology itself in the new ASR actor-network for the Court will be problematic, due to the errorful, disobedient, nature of ASR software. The main technical difficulties associated with using ASR in the courtroom are:

- Less than perfect recognition accuracy;

- Poor acoustics in the courtroom;

- The large amount of out-of-vocabulary (OOV) words used in the process; and,

- The lack of a clear way to map what a magistrate says to what is recorded in the Courts' computer system.

Automatic speech recognition technology is currently far from perfect. In addition to the low recognition accuracy, the errors ASR makes are not comparable to those that a human makes. It is possible that if an ASR system makes a mistake it will incorrectly transcribe numbers and names, two elements that are vital to the accurate recording of an outcome in the courtroom. There are two ways in which it is common for an ASR system to fail in correctly recognising a word. The first is a recognition error and the second is an out-of-vocabulary error.

In a recognition error, a word is misrecognised ("Australians" transcribed as "astray aliens", for example) due to the combination of the speaker's accent and the way in which the software recognises words. These sort of errors can generally be trained out of the speaker-speech recognition software pair over time as the software learns the speaker's voice and the speaker adapts to the software. A second kind of recognition error is context error. In a context error a word is incorrectly transcribed due to the software's lack of context awareness. A simple example of a context error is the substitution of the word "carrot" for the less common word "caret". Adapting ASR to the Court would require the creation of Court-specific vocabulary and acoustic models.

The obdurate elements in the heterogeneous assemblage that is the court are:

- The social world of the Court;

- The Court room layout, as it influences the social world of the Court;

- The work process in the courtroom and all public-facing areas of the Court; and,

- The requirement to record decisions on outcomes made by the magistrate during court.

The elements in the Court that are more plastic are:

- The detail of the work process of the "back room", particularly the after-court section; and,

- Details of the defendant's folder but not its use or existence.

7.2.3 Points of Passage and Trials of Strength

In Actor-Network Theory, a "point of passage" is an actor who directs, or tries to direct, the network to their interests. In the case of the new network at the ACT Magistrates Court, the "point of passage" whose interest is of importance to a technologist is ASR itself. Other actors in the new network may attempt to assert their power as points of passage but the power of the Chief Magistrate in the Court would seem to be so much greater than that held by the other individuals in the Court's work process that any objections by them would be ignored or cast aside by the Chief Magistrate who is seemingly in favour of the ASR system.

A "trial of strength" is when all the heterogeneous elements of an actor-network must perform their roles. Discussing the trial of strength of an imaginary system is largely moot, but by speculating about such an event points of weakness can be identified and later strengthened. The most obvious point of technical failure is the ASR system itself, not just within the recognition software but the wider system of vocabulary-models, acoustic models, microphones, cabling, user-interfaces and so on. Careful system design involving the users of the system in a user-centred process would lessen the severity of this point of failure, particularly if an iterative user-centred software development process was followed.

The non-technical points of possible failure are more difficult to "design out" of concern because they are less predictable. The best way to "design out" the non-technical points of possible failure is to "design them in" by respecting and valuing the existing work of the human actors in the system and using a new design to aid them in their skilled work. This "designing in" (ideally) takes place during the Enrollment stage and is necessarily a process of negotiation.

7.3 Re-imagining Automatic Speech Recognition

To use ASR in the ACT Magistrates Court necessitates that ASR, as a technology, be re-imagined. Often, ASR applications are seen as being a replacement for typing to be used by one user, that is, the dictation paradigm. In the dictation paradigm, an ASR application is used to replace a secretary who takes dictation as the user speaks. However, this is not the only paradigm for the use of ASR.

The re-imagined form of ASR that would work for the Court would not use the one-user-to-one-computer model of dictation but a model where the users and computers are distributed in space and time as the work process of the Court is distributed in space and time. Inherent in this distributed model is the fact that the person whose speech is recognised is not necessarily the person working with the transcript generated by the ASR system. Distributing the computers involved allows separation of work tasks and recognition tasks as well as allowing multi-pass ASR (Whittaker et al., 2002) which can improve the accuracy of hard-to-recognise speech by allowing a recogniser to refine a transcription.

As stated in the previous section, the elements of the Court that are most plastic, and therefore easiest to change, are:

- The detail of the work process of the "back room", particularly the after-court section; and,

- Details of the defendant's folder but not its use or existence.

This is not to say that these elements will be easy to change, just that they are easier to change than, say, the physical layout of the courtroom. Analysing the work of the Court has shown that these elements are the most flexible to change and that is where the design work is concentrated. Having identified the following elements that are particularly resistant to change this design does not attempt to encroach on their existence, though it will necessarily have follow-on effects that cannot be predicted. These resistant elements are:

- The social world of the Court;

- The "theatre" of the courtroom;

- The Court room layout, as it influences the social world of the Court;

- The work process in courtroom and all public-facing areas of the Court; and,

- The requirement to record decisions on outcomes made by the magistrate during court.

Re-imagined ASR for the Court incorporates a model of the Court's work-flow. In the current work process the magistrate speaks a decision and writes it down and other people perform the coding of that decision into something that allows the machinery of the Court to keep working. A dictation paradigm of ASR can't perform that task because the translation of the magistrate's decision into codes is too nuanced, too detailed, too specialized and too reliant on intelligence and experience. The goal of a re-imagined ASR is to make it easier for the back-room workers to do the parts of their job where intelligence is required and minimise the parts of their jobs that are repetitive.

The Chief Magistrate's request was that any new system remove the need for him to write down every decision. It is quite simple, technically, to record the speech of the Chief Magistrate when he makes a decision. In fact, it is done already and annotated by the monitor. However, using those recordings as a resource to replace or augment the bench sheet is considerably more problematic. Adding to the possible problems of moving to an speech record is that the stamps which are currently used and which may act as a prompt to the magistrate are no longer available. Minimising the amount of writing on a bench sheet by a magistrate requires that the information previously contained on the bench sheet is available elsewhere. Given that the magistrate's speech is recorded it is reasonable to attempt to provide access to that recorded speech.

There is already a body of work on accessing speech recordings (see, for example, Whittaker et al., 2002). The problem with accessing speech recordings for the Court, and the back room in particular, is that using the bench sheets and the entire folder is a fact finding exercise where users scan and browse the documents in the folder looking for the specific information that the case-management software requires. Replacing the bench sheet, which at least in part embodies the process of communicating decisions, with an ASR system is a similar problem to that faced by the designers of the air traffic control system described by Hughes et al. (1992a) where a design for replacing paper that was an embodiment of work was proposed. As with the air traffic control work by Hughes et al. the proposed design for an ASR system for the Court attempts to retain the communicative aspects of the embodied work while introducing new possibilities for interaction to the process.

Using a speech record instead of paper to communicate decisions necessitates scanning and browsing a recording of the magistrate's speech in court. Scanning and browsing a recording of speech is time consuming because speech is one-dimensional and ephemeral. One way of making speech persistent and two-dimensional to support scanning and browsing behaviours is to turn speech into text. A group at AT&T Research looked at voicemail and speech in general and developed the Spoken Content-based Audio Navigation (SCAN) interface. SCAN was developed to be used as a way to access transcripts of broadcast news (Whittaker et al., 1999) and later voicemail (Whittaker et al., 2002). The AT&T researchers had no illusions about the errorful nature of ASR, however they showed that the errorful transcripts allowed users to obtain an overview of audio recordings that was previously impossible. The errorful transcripts turned unusable speech recordings into something useful.

It is important to note that the interface described below is not a proposed solution, but is a way of exploring and building on the ideas presented until this point in this thesis. A solution would need extensive testing with proposed users and would have to undergo several iterations before it would be ready to be used "live". The design for the interface is speculative and arises from the fieldwork described elsewhere in this thesis. Presenting this design here is not an attempt to say "here is an ideal design for ASR in the Court" but rather a way to show how the fieldwork has led to a design outcome.

The interface described in the next section has some similarities and some differences with the SCAN (Whittaker et al., 1999, 2002) interface. The first use of SCAN was as an interface to archived broadcast news recordings and was intended to solve what the researchers termed the under-utilisation of growing archives of speech collected from radio programmes, Congressional Debates and private archives of audio conferences. SCAN was the implementation of a new paradigm in accessing speech records, What You See Is (Almost) What You Hear or WYSIAWYH. The primary goal of WYSIAWYH was to present a visual analogue to recordings of speech. SCAN used transcripts of speech generated by ASR software to facilitate the visual analogue. To create the transcripts the speech was broken into "paratones"[5] and then passed through an ASR engine several times, allowing the recogniser to improve on the transcript. The results of the transcription for each paratone was then combined into the errorful transcript of the particular audio recording. The terms in the transcripts were then indexed for later retrieval. Users could enter natural language queries into the SCAN interface and the system would return ranked transcripts that the user could select to view and, if required, listen.

SCAN had an "overview" feature that displayed the incidence of keywords in the paratones of the transcript and the transcript itself. By providing a visual overview of the keywords in various paratones, SCAN allowed the user to skim the document more quickly than if they had to scan the entire transcript, which could be the textual representation of 25 minutes of speech. After using the overview section to jump to a potentially relevant paratone, the user could read the (errorful) transcript. If the transcript contained too many errors to be sensible the user could click the paragraph to play the audio it represented.

The SCAN interface was empirically tested and found to be more effective than just listening to the recordings in fact-finding tasks. Additionally, subjects in the experiments found the SCAN interface easier to use than just listening to the audio and listened to a shorter amount of audio to complete the tasks set them. The researchers found that increases in transcript accuracy had an influence on the perception of the difficulty of the task and on the actual quality of the answers the subjects gave. The mean accuracy of the transcripts in the tests was 67% with a maximum of 88% and a minimum of 35%. SCAN was found to be particularly useful for fact-finding tasks using the broadcast news corpus.

A second report on the SCAN interface introduced a voicemail corpus and again tested the interface with users (Whittaker et al., 2002). In this experiment, the interface was called SCANMail. The interface for SCAN-Mail resembled a typical three-pane e-mail screen with one exception, it also contained a simple audio player interface and a thumbnail represen-

---

[5]Paratone is Whittaker et al.'s term for "audio paragraphs".

tation of the transcript of the voicemail. The overview feature of the previous SCAN interface was not used. The SCANMail browser featured new
behaviours, too. In addition to being able to play the audio behind paratones, users could select regions of text within paratones, or even over
paratones, and play the underlying audio. The system also attempted
to extract relevant information in the voicemails, particularly telephone
numbers, which were represented in the transcript with icons. As with the
previous incarnation of SCAN, the SCANMail interface facilitated search-
ing the entire corpus of voicemail. Upon doing a search, the keywords
were highlighted in the transcript and in the transcript thumbnail.

The researchers note that there are disadvantages to the SCAN approach. The chief disadvantage is over-reliance by users on transcripts.
Because the transcripts are inherently errorful relying on them can introduce errors into information extracted from the transcripts. They suggest
introducing a representation of the ASR confidence measure, that is word
probability, to the transcript. Words that the system was more confident
about could be presented in darker type and words that had a lower confidence could be presented in a lighter type allowing the user to judge
for themselves the accuracy of the text. Some users in the studies of the
SCANMail interface suggested that the transcripts could be editable to allow for correction of errors should they be found. Whittaker and Amento
(2004) built and tested an editable version of the SCANMail interface and
found it to be usable and useful.

For clarity, it must be said that the interface described here has no relationship to the caricatured interfaces described in sections 7.1.2 and 7.1.3.
I will refer to the interface described in this section as the *Interface for Court
Audio Access* (ICAA). The main difference between SCAN and ICAA is
that SCAN was intended to provide open-ended search capabilities over
a large corpus of speech, either broadcast news (Whittaker et al., 1999) or
voicemail (Whittaker et al., 2002) where ICAA would not require the ability to search over all speech recorded by the system but would instead be
directed at searches of a single transcript or group of transcripts relevant to
a particular case. ICAA would replace bench sheets or augment a greatly
simplified version of the existing bench sheets, allowing the magistrates

freedom from writing large amounts by hand while still allowing workers in the back room access to the information they require to perform their work.

In the previous chapter, a trajectory of using ASR in the workplace was described. Within the trajectory are nine points that were identified as essential for using ASR. This trajectory is:

- Deciding to use

    1. Reasons for use

    2. Previous experience with similar technology

    3. I.T. savvy

- Leaping into use

    4. Training

    5. Hardware and software

    6. Fit with work

- Sustaining use

    7. Necessity of use

    8. Technical support

    9. Becoming the conductor

The points in the trajectory are addressed in the design of ASR for the Court which is described in the next section.

7.3.1   Scenario: Interface for Court Audio Access

The Interface for Court Audio Access (ICAA) scenario is partly inspired by the work of Kraal et al. (2002). In this section, the scenario itself is presented in a sans-serif typeface with occasional explanations of action in the regular document typeface.

In keeping with previous work that has used ethnographic methods (Hughes et al., 1992a) and scenarios (Satchell, 2003) to describe future designs, the technical details of the scenario that follows are not described. The purpose of the scenario is to describe the system *in use*.

In this scenario, the **Deciding to Use** part of the trajectory has been passed. The ACT Magistrates Court's *Reason* for using ASR is that it has decided to do away with the magistrate's handwritten bench sheet. The Court has *Previous experience* with the magistrate speaking sentences aloud and with making audio recordings of what is said in court. The Court is *Savvy* enough to be aware of ASR in order to re-imagine the back-room process so that it fits with the courtroom use of ASR.

It's Monday morning, always the busiest time for the A-list with all of the weekend arrests to deal with, and Court has just resumed at 11.07am, Magistrate Rob Cowley presiding. They're up to the drink-driving charges. First up, Henry Webb, representing himself. Claire hands up Mr Webb's folder. As it crosses the boundary from Claire's desk to the Bench, the touch-screen on the bench shows the charge numbers for the case in the folder—Mr Webb's driving under the influence charge—there's only one number. Mr Webb pleads guilty but states that this is his first charge for driving under the influence in 38 years of driving and indeed his first criminal charge ever.

Rob asks the public prosecutor what Mr Webb's blood-alcohol content was. "Zero point zero six, your worship". Barely over the legal limit and fairly obviously a lapse of judgment on Mr Webb's part. Rob notes it down on a blank sheet of paper in the folder in front of him. He's obviously contrite and just appearing in court seems to have scared him so much he'll be catching cabs from now on. Rob decides to give Mr Webb a good behaviour bond and a stern lecture.

Use of ICAA begins in the courtroom when no actual "interface" is visible. ICAA's intrusion into the courtroom itself is limited to a microphone, a few RFID[6] sensors, a small touch screen on the bench and a small, fast, printer on the associate's desk. Here the court has **Leapt into Use**. The *Hardware and Software* are in place and the system *Fits with the Work* be-

---

[6]Radio Frequency Identification.

ing done. Though the system is fairly unobtrusive, the magistrate and the associate have been *Trained* in the use of the system.

As court progresses, ICAA makes no intrusion into proceedings until a case comes to a point where the magistrate would previously have written a decision on the bench sheet.

The microphone and touch-screen are directly related to ASR. The magistrate uses the touch-screen as a way to start and stop the speech recognition when he's speaking a decision.

The printer on the associate's desk produces dockets that show a decision, or series of decisions, have been made relating to the case at hand. The RFID sensors sense RFID tags embedded in the folder. As the folder is passed between associate and magistrate sensors in the bench record the passing, allowing the system to dip into a database for pertinent information, for example charge numbers and various details relating to the defendant, if known, such as address and employer. The touch screen can then display these details.

"... use better judgment in the future, won't you."

"Yes, your worship."

Stern lecture over, it's time to sentence Mr Webb to good behaviour. Rob taps the touch-screen to start the decision-recording process. The gesture is so subtle that no-one in court really notices it. The screen shows READY FOR DECISION and still shows the charge numbers.

An audio recording has been going on since Rob sat down and court began. When Rob taps the screen to tell it he is about to speak a decision, the system tags the recording, allowing a future listener, or the ASR system, to jump to the sentence.

"In the matter of charge number HW39674, Henry Webb is hereby released on recognisance self in the amount of $1000 on the condition that he be of good behaviour for twelve months."

Rob taps the screen again, ending the recording. The screen shows RECORDING FINISHED. Rob hands Mr Webb's folder back to Claire and as it crosses the boundary from the bench to her desk the touch screen shows NEXT CASE. At the same time, a small printer on Claire's desk produces a docket with a ten-digit number and a few details relating to the case. She puts it in the folder and puts the folder on her "done" pile.

Mr Webb's day in court is over and he's free to go.

So far, most aspects of the court's work process are much the same as they are currently. Handwritten decisions have been done away with, as was the purpose of this design, and replaced with what is from the magistrate and associate's perspective technology that is unobtrusive. The technology introduced into the court is strong and simple, in keeping with the findings that the ASR system introduced to the Court and does not significantly disrupt the work process in the courtroom or impinge on the theatre of the court.

While Mr Webb has been getting his lecture, and indeed since court has started, Molly has been in the monitor's booth watching and listening to everything. Molly has a computer in front of her with special software that can annotate the audio recording of what's going on in court. Since this is the A-list, Molly's job is just to record which lawyers are appearing when. Molly also has a paper master charge sheet listing every charge that's appearing in court today. She uses the charge sheet to record which charge numbers are dismissed and which charge numbers the magistrate decides to deal with.

In theory, with the ASR system in place, the monitor's job is unnecessary[7], however, ICAA keeps the monitor's job and makes the annotations work as part of the ASR system.

The monitor still uses the paper charge sheet to cross off dismissed charges so that the ASR system has a back-up in case something goes very wrong. The charge sheet also helps the person doing the after-court processing, the process of which is described later in this scenario.

The monitor has also **Leapt into Use**. Molly has been *Trained*, she has *Hardware and Software* that allow her to use the part of the ICAA that she needs and the system *Fits* with her work.

The next part of the ICAA scenario takes place in the "back room" of the Court, when all of the defendants' folders are being processed by Car-

---

[7]Indeed, the ACT Magistrates Court is the only Magistrates Court in Australia that has a monitor, or monitor-like position. Anecdotally, the reason for this is that that Court is much better funded than other Magistrates Courts.

men in the after-court section.

The defendants' folders and the monitor's master charge sheet make their way to the back room and become the responsibility of Julie. Julie works in the after court section, processing folders from the day in court and entering details of the magistrates' decisions into the Court's case management software. The ICAA and the case management software (CMS) work together to help Julie do her job.

Julie takes the first folder, which belongs to a Mr Smith, from the big pile next to her desk, opens it and types the code on the docket at the top of the documents in the file into the ICAA. This works much better than the way things were about a month ago when they installed sensors in Julie's desk to automatically detect which folder Julie had selected. The sensors worked fine but they meant that Julie couldn't place the folders on her desk the way that she used to. Julie had the I.T. guys remove the sensors—she's happy to type a number if it means she can put the folder she's working on wherever she likes.

The folders and the RFID sensors work in the courtroom because there is a very clear demarcation between the bench, where the folders are "in play" and the associate's desk where the folders are "waiting". On Julie's desk the distance between the "in play" area and the "waiting" area is too fine and too variable for the sensors to work reliably. For Julie, the sensors got in the way, though for the magistrate and the associate the sensors are helpful.

Julie had the ICAA thrust upon her by the court **Deciding to Use** it. The ICAA is intended to help the magistrates in court and allow them to not write down decisions on outcomes but it still needs to communicate those outcomes to the people who need to know them. Julie's *Reason to Use* her part of the ICAA is that it is her job to process what is said in court. The ICAA is designed to make it as easy as possible for Julie to work with the audio and the transcript that is generated from it using ASR. Julie's *Previous Experience* with her job before the ICAA was introduced allows her to work with the transcript. Finally, Julie is *Savvy* enough to understand that the transcript is pretty close to the old handwritten bench sheet. She could either fight the new system or see it as a different skill to learn.

After entering the code from the docket, the ICAA case window appears with the most recent transcript from Mr Smith's trial already open in the transcript pane. If there were other transcripts from previous appearances, they'd be in the archive pane, but this is Mr Smith's first time in court. By scanning the transcript, Julie is able to assess what has happened in court and what decisions the magistrate has made. In this case, Mr Cowley has dismissed a bunch of charges and set aside hearing the remaining charges for a later date. Clearly this person has pleaded not guilty. The ICAA is really good at recognising charge numbers so Julie quickly scans the transcript to make sure that nothing is really wrong and tells the ICAA to tell the CMS to record that the charges were dismissed. All this takes is a few mouse clicks.

The ICAA is so good at recognising charge numbers because the touch-screen shows the magistrate the charge numbers for the case at hand. This serves two purposes. It prompts the magistrate so that the charge numbers are easy to view and it primes the ASR software so that when it "hears" a charge number it will only recognise it if the number is from the list of charges in the case at hand. The RFID tags in the folder allow the ASR system to narrow the possibilities of what charge numbers the magistrate could say, leading to better recognition accuracy.

After taking care of the dismissed charges, Julie is able to get the longer part of Mr Cowley's decision where the case is set over for a date in three weeks time. The system has jumped through the transcript to the next part of the decision. Mr Cowley said that he'll hear the case on the 23rd of this month. The system understood that really well as it's in black text. He gave a few other orders that the system isn't that confident it's understood—they're in varying shades of gray—though they make enough sense as Julie reads through the transcript.

Using different shades to display the confidence of a recognised word or phrase has been used successfully in other transcript-based interfaces to underlying audio (Whittaker and Amento, 2004).

Julie is able to select the part of the transcript that has the date in it and drag it to the field in the CMS that accepts dates. The ICAA knows that the

CMS wants dates in a YYYYMMDD format and can convert "23 January" into 20060123 on the fly. Julie makes sure the conversion is correct. Now she switches her attention to the CMS pane and fills in the rest of the required information. Mr Cowley has neglected to say which charges he'll be hearing on the 23rd, which isn't a problem in court as it's fairly obvious when he's dismissed a lot of charges, but the CMS needs to know exactly which ones he'll be hearing. The CMS assumes that unless charges are dismissed they're still current, so Julie confirms that with the CMS and checks quickly with the master charge sheet from the monitor. Before this folder is done, Julie has to print the CMS's summary of the outcomes so far and some letters to send to the various parties involved in the case. These letters are just proforma and are generated by the CMS. A letter for the public prosecutor's office; one for Mr Smith; one for Mr Smith's lawyer. They're printed in duplicate; one copy for the folder and one copy for Julie's outbox. While the printer takes its time, Julie pulls out the next folder, Ms Barker.

The next folder is quite thick. Ms Barker has generated a lot of paperwork and has obviously been in court many times. Since this is the A-list pile she has probably re-offended while on bail. Julie quickly types in the code number from the docket from the top of the folder. She sees that the system has not managed to make a very good transcription. Bad transcripts are always different and this one starts, "butler company on does enter..." all in black. It's weird how sometimes the speech recognition can be confident about gibberish and not confident when the transcript makes perfect sense.

Confidence measures indicated by the blackness of the text in ICAA don't indicate semantic correctness, simply algorithmic confidence. Julie and the other people who use transcripts from the ICAA are trained so that they know the difference.

Julie and her colleagues have **Leapt into Using** the ICAA. They've been *Trained* to understand the sometimes strange things that ASR software can do. They've got the *Hardware and Software* they need to access transcripts and the underlying audio. And the design of the back-room part of ICAA *Fits* with how they work, looking for specific words and meanings in records of outcomes.

Scrolling down shows that the rest of the transcript is not much better. Selecting the first paratone in the transcript, Julie plays the audio, "But her companion doesn't..." - ah that explains it. The magistrate has woken up ICAA in the middle of speaking which always seems to confuse it. No matter as the audio is good, so Julie can listen to the judgment. This time it is an order to undergo counseling and drug rehabilitation at a facility 300km to the east. The system invariably gets the name of that facility wrong in a transcript anyway, so Julie resigns herself to the fact that she would have had to listen in even if the transcript was good. While she listens to the rest of the audio, Julie picks up the letters from the printer and files them appropriately, distributing them between Mr Smith's folder and her outbox. Switching her attention to the case management software, Julie checks that she is looking at the relevant case and charge (there's only one) and enters the information by hand. This requires more letters be printed. While the printer whirs away at these, Julie picks up the next folder.

The Court, in this scenario is **Sustaining its Use** of the ICAA. It is completely *Necessary* that they use ICAA as they have no other way of communicating outcomes from the courtroom to the back-room. They have *Technical Support* (though it is not described in this Scenario) that ensures that the hardware and software interfaces work as they should. Finally, the court organisation, through its commitment to the ICAA, acts as the *Conductor*, ensuring that every aspect of the ICAA continues to work in harmony.

CHAPTER 8

Reflections on Methodology

This chapter reflects on the methods used in this work that led to the design of an automatic speech recognition (ASR) system for the ACT Magistrates Court. The methods used in this work were treated as "tools to think with" and were always used with designing in mind. This chapter reflects on the use of these tools and the thinking they inspired to answer the research questions:

1. How is ASR used and made usable and useful in the workplace now?

2. How could ASR be used and made usable and useful in workplaces in the future?

To answer the first question it was necessary to undertake fieldwork which involved going to places where people use ASR systems and observing and interviewing them. These observations and interviews generated rich data that was then analysed using two socially-oriented methods, the Locales Framework (Locales) and Actor-Network Theory (ANT).

Answering the second question involved investigating a workplace where ASR could be used, the ACT Magistrates Court, and designing an ASR system for that workplace. Investigating the Court generated rich data in the form of observations and interviews which was analysed using Locales and ANT.

The analysis of the Court's existing work processes and the ASR users was then used to inspire the design for ASR at the Court.

The methodology used for moving from fieldwork to design in this thesis aligns quite closely with "Technomethodology" (Button and Dourish, 1996; Dourish and Button, 1998; Crabtree, 2004). Crabtree (2004) describes technomethodology as having a hybrid methodology:

1. Let designers build whatever they want with whomever they want.

2. Deploy the objects of design in real world settings.

3. Treat deployment as a breaching experiment.

4. Explicate the accountable structures of practical action made visible in the breach.

5. Explore the topics identified in the breach through the study of perspicuous[1] settings.

6. Use the studies of perspicuous settings to flesh out abstract design concepts and work up design solutions in conjunction with the other parties to design.

7. Deploy the new design solution in real world settings and study its use.

8. Repeat the process until the research agenda has been satisfied for all practical purposes.

In this thesis I have broadly followed steps 1 through 6 in Crabtree's hybrid methodology with some modifications. In particular, I have compressed Crabtree's steps 3, 4 and 5 into the one process.

Step 1. Crabtree says to allow designers to build whatever they want. In this work I have interpreted this to mean investigating commercial-off-the-shelf software.

Step 2. Since the software identified in step 1 is commercial, it is readily found in real-world settings. The real-world settings in this thesis are various offices in the public service and Hansard.

---

[1]Perspicuous, or perspicacious, or clear, or intelligible.

Steps 3, 4 & 5.   A breaching experiment is an experiment that attempts to provoke a reaction by violating commonly accepted ways of doing things. In this work, the use of ASR systems in the workplace is a breaching experiment because speaking to a computer is not a normal way of working.

As Crabtree says, "The emphasis placed on disrupting everyday activities is overstated, however, and even misleading if taken too literally." This is particularly true of the situations I have examined in my thesis where people are using ASR software because for them, the everyday activity *is* using the technology. Nevertheless, getting everyday activities done with ASR provides many instances of natural breaches. At the same time, for those who must support the ASR users, the ASR software is disruptive to their everyday experience and this is revealed in the difficulties that the ASR users have in enrolling coworkers to assist them when they encounter difficulties with the ASR software.

In keeping with Crabtree's step 4, I have analysed how people work with ASR day-to-day by exploring the "practical actions" they take to allow themselves to keep working. This has taken the form of examining the practical action that users take in order to sustain their use of the ASR software. The detail of actual practice can help the designer of a future system because it can reveal where a user has to work around the existing design.

I have interpreted Crabtree's step 5 directive to "explore the concepts identified in the breach in perspicuous settings" as a way of looking at the use of ASR in the workplace by treating breaching experiments as perspicuous settings where it is convenient to do so. This re-interpretation is done through the analysis of the fieldwork using Locals and ANT which are used to *make clear* what may otherwise be seen as a breach. Because the "breach" and "perspicuous setting" are both examples of the same lived practice the data is representative of the real-world. The re-interpretation of this step and the re-interpretation of step 6 are the greatest departure in this thesis from technomethodology.

Step 6.   In this step, Crabtree expects that a process of user-centered collaborative design will take place. As it was impractical within the scope and resources of this research to build and implement an ASR system for the ACT Magistrates Court, this thesis instead presents an analysis of the sentencing work process and a proposed design for the Court.

Steps 7 & 8   These steps were not used in this thesis.

Crabtree expects that the artefacts built in step 1 will be built by designers who are involved in the larger project. In my research I have used pre-existing commercial products as the jumping-off-point for my investigations. This has allowed the short-circuiting of step 2 as the artefacts were already being used in real-world settings. Treating the use of the artefacts as breaching experiments then became one of perspective rather than intent. Similarly, treating the artefacts as perspicuous settings is also one of perspective.

Technomethodology, in Crabtree's version, requires novel designs to be created before the process of investigating them in the real-world can begin. In the methodology followed in this thesis:

1. A novel way of working at an existing location is postulated (ASR at the Court).

2. Contemporary versions of the future way of working are identified and studied for their essential properties (ASR users at Hansard and in the Public Service). These may be treated as breaching experiments.

3. The work practices of the existing situation are examined and analysed, making the "practical actions" clear and allowing the work to be treated as a "perspicuous setting".

4. The outcomes of step 2 and step 3 are treated as input to a design process.

5. A novel design emerges.

In this methodology, the use of ASR applications has the qualities of breaching experiments in that it disrupts the everyday activities of the user and those around them. It forces the user to change their way of working and it can force people who support the user to change their way of working as when some of my subjects changed jobs to allow them to use the software. The introduction of ASR also impacts on how users interact with other software, at times forcing them to adapt the ASR software to the new software, as when my subjects describe writing "macros" or it can even influence the choice of new software as when Susan said that she was included in the decision-making process for her office's new document management software to ensure interoperability.

The use of ASR applications in the workplace acts as a perspicuous setting because practice is already somewhat established—instead of "calling forth" the practice, as in a breach, I have tried to make the practice clear by raising it into view through analysis.

By investigating ASR applications "in the wild" I have been able to see, and show, the social circumstances that make the applications useful. In turn, I have then made the move into *beginning* design informed by the social findings of the investigation. In performing this methodology I have used Crabtree's research and development model to see technology as a vehicle for social research and used the results of that research to propel innovation and design. A socially oriented research methodology has been particularly useful for investigating, and designing for, automatic speech recognition.

8.1 Social Research for Automatic Speech Recognition Design

Designing a usable ASR system is often presented as being about getting the grammar and vocabulary of the system right and achieving as low a recognition rate as possible (Huang et al., 1999; Weinschenk and Barker, 2000) despite some early fieldwork indicating that a close fit with the task at hand is important (Rollins et al., 1983). Van Buskirk and LaLomia (1995) showed that the "just noticeable difference" (JND) in speech recognition accuracy is somewhere between five and ten percent. This means that a user cannot tell the difference between, for example, 90% accuracy and

97%. Since most ASR researchers would agree that 100% recognition accuracy is impossible at current levels of technology, the JND of between five and ten percent implies that once the recognition rate is high enough other factors become more important with regard to the usability of a system. As I have shown in Chapter 6, the usability of ASR systems in the workplace is highly dependant on many factors external to the software interface. Thus, when considering a new ASR application it is important that these non-recognition-accuracy factors be taken into account. This work has taken the view that the recognition rate of commercial ASR software is sufficient for workplace use and that opportunities for improving the usability of ASR software are to be found in bringing the capabilities of ASR software closer to the tasks that (potential) users perform daily. The methodologies used in this work have therefore been directed at discovering and understanding those tasks and how they are achieved.

In examining users of ASR systems, the most critical non-recognition-accuracy factor in the successful use of an ASR system is that of *fit with work*. Fit with work is the degree to which it is possible to perform the task at hand given the capabilities of the ASR software. If the fit with work is low and it needs to be higher, the capabilities of the ASR software can be changed, the way of completing the task can be modified, both can change or the goal of completing the task with an ASR interface can be abandoned. Because there is limited scope to change the capabilities of commercial-off-the-shelf ASR software, contemporary users of ASR software often find it easier to change their manner of task completion or abandon trying to complete the task with ASR altogether. This typically takes the form of moving to a job where the tasks are better suited to the capabilities of the software.

When considering a new ASR application it is important to research in the workplace where the new application will be used to find the best fit with the capabilities of ASR software. Exploring the social situation and gaining an understanding of the degree to which the workplace could support the critical non-recognition-accuracy factors is also important. In this thesis the methods for performing the social analysis required to understand the work situations of the ASR users and the ACT Magistrates

Court were the Locales Framework and Actor-Network Theory.

8.2 Using the Locales Framework

The Locales Framework presents a comprehensive approach to analysing qualitative data. The extent of Locales is daunting, encompassing almost every aspect of cooperative work. Fitzpatrick's own use of Locales is selective (Fitzpatrick, 2003), as is that of others (Graham et al., 2005), showing that Locales is a tool for appropriating in use, rather than a specific method to follow exactly. As such, the use of Locales for analysing qualitative data in this thesis has been selective. It has been used to ask questions and to answer them.

Locales has been used to ask questions about the qualitative data obtained through interviews and observations. Using Locales as a "sensitising" tool was useful in directing the research towards aspects of the data that were particularly interesting, for example the embodiment of the sentencing process in the artefact of the defendant's folder.

Locales has been used to answer questions that emerged from the qualitative data. The questions that emerged from the qualitative data were comparative, e.g.: How is the situation of the injured users similar to that of the Hansard users?

By using Locales to ask and answer questions, I have used it as a corpus of social knowledge relating specifically to the use of technology in work. Locales can be used this way as it is intended to make social thinking accessible for technologists. Where concepts from Locales were particularly useful, they were followed up in source literature, thereby expanding the detail of the corpus used to perform the analysis. Being able to analyse fieldwork data in this manner, with reference to a corpus of existing analyses was valuable because it allowed more confidence in making conclusions that were reflected in the work of others and it placed this work into a wider sphere of similar work.

Using Locales to design is primarily an exercise in exploring ways to enhance locales by using existing spaces and resources differently or by evolving entirely new locales (Fitzpatrick, 2003, p.g. 149). Because each design is necessarily unique to the locale in which it is situated, Locales offers no design advice, instead acting to direct the designer to support human action and interaction. Fitzpatrick says that the use of Locales for design is "driven by interactional needs and understanding the broader context(s) in which those interactions happen, not by *a priori* assumptions of technological solutions" (Fitzpatrick, 2003, p.g. 149). In this thesis, the use of ASR was assumed *a priori* though there was no assumption about the form that the ASR would take. Instead, this thesis re-imagined ASR in order to fit with the work process at the ACT Magistrates Court. This *a posteriori* justification for imagining the use of ASR at the Court brings the process back to Locales by showing that the use of ASR could lead to new ways of working at the Court. Further, by engaging in technology-driven (Danis and Karat, 1995) speculative design for ASR, new ways of imagining ASR have emerged from the design process.

Where Locales has proved most useful in designing for the Court was to act as a brake on my enthusiasm for exciting possibilities that were otherwise hardly considered and potentially difficult to use. When designing, as when analysing the qualitative data, Locales was used to ask questions of the design as it evolved.

Fitzpatrick says, of using Locales in design, "The types of questions [...] would be domain specific, and guided by the purposes for engaging in this exercise in the first place" (Fitzpatrick, 2003, p.g. 149). As has already been said, the reason for engaging in design for the Court was to explore how ASR technology could be used while still respecting the existing work process. The questions asked during the design phase were to understand the work process so that it could be respected. For example, because the locale of the courtroom is so rich and relies on theatric aspects to maintain the authority of the magistrate, every decision to introduce technology to the bench was questioned: Is this necessary? Is there another way? Can this be achieved more "calmly" and "quietly"? Questions were also asked when technology was introduced to the "back room" at the Court: Will this disrupt the existing work? Does this make the work more difficult? Is the new way better than the old?

8.3 Using Actor-Network Theory

Actor-Network Theory (ANT) is an approach to doing sociological analysis. ANT comes from the area of sociology called "sociology of science and technology" (STS) which has tended to be interested in how science is done and how the social influences science (Latour, 1987). Increasingly, ANT is being used in computer science and information systems as a way to analyse how computer systems are used in the workplace (Tatnall and Gilding, 1999; Abdelnour Nocera and Hall, 2004). Using ANT for systems analysis of computer systems is well within its capabilities. Like Locales, ANT can be used piecemeal or in a minimal fashion, though, unlike Locales I have found that it is necessary to engage with ANT as a whole before it was possible to determine which pieces to use in this work.

Of particular inspiration to the use of ANT in this work was John Law's 2003 description of ANT as talking about representation in terms of translation. ANT tells stories about how things, objects, actors, come to be how they are. In ANT, actors come to be how they are through a process of interaction with other actors. Interaction changes actors. It translates actors. The idea of translation through interaction resonates with the experience of using dictation software and making it useful in the workplace. In using dictation software in the workplace, the user, the software and their work are translated. The user becomes a person who knows about the nuances of dictation software—they become an expert. The software is changed (usually) in order to become ever-so-slightly more closely aligned with the user's work. In some cases the user's work is changed, sometimes dramatically, in order to better fit with what is possible to achieve in using the software. The user, the software and their work are no longer the same after an interaction. They are translated. In technomethodological terms, it is possible to explain the translation as a breaching experiment and then use that explanation as inspiration for design.

Law (2003) describes Akrich's work on machines that made briquettes for use in fireplaces. The machines are originally made in Sweden and are then sent to be used in Nicaragua. The way the machines are used in Sweden and Nicaragua are so different as to be almost unrecognisable. The way the machines came to be so different, to be translated so significantly, was through a series of negotiations. In this work, the translation of ASR software is not so significant, yet in comparing how dictation software is used at Hansard and by injured users elsewhere in the Public Service it is also possible to show that they are different due to a series of negotiations.

"The two networks are different in every respect" (Law, 2003). This is true in Akrich's work as it is in my ASR users study. The injured users' and the Hansard users' networks look very different, despite using the same software for ostensibly the same purpose. As the technology is re-made and re-negotiated for each situation it is translated by the users, the organisation and even the work situation. It even plays different roles in each situation; for the Hansard users it plays the role of a tool in an array of tools, to be used when necessary, but for the injured users it plays the role of a limb, to be constantly acted through and with.

Also shown in my work is the idea that enrollment is precarious and requires constant maintenance so that the links in the network are sustained. A good example is Wendy who was the victim of the links in her network failing compared with some of the other users who were able to maintain their networks. By maintaining their networks, or at least by not having them crumble, most of the users were able to establish their practice of using dictation software. Once practice is established, it becomes easier to sustain the existing networks. Practice stabilises the networks.

ANT is concerned with *how* translations happen and how social order is established and maintained. For my purposes, what is meant by social order is the consistent use of software in a working environment. Akrich is particularly concerned with the *obdurance* or plasticity or malleableness of a technological artefact. In Akrich's terms, *adjustment* is what happens when people and technologies work together; when people adapt to technologies and when technologies are changed by people.

> "If we want to describe the elementary mechanisms of adjustment, we have to find circumstances in which the inside and the outside of objects are not well matched. We need to find disagreement, negotiation, and the potential for breakdown." (Akrich, 1992)

These terms, looking for disagreement, negotiation and breakdown, have resonances with breaching experiments. A breaching experiment is intended to force some new practice into the open by using the technology in the world so that "The world inscribed in the object and the world described by its displacement" (Akrich, 1992) can be lifted up for viewing by designers. Designers can then inscribe "the world described by its displacement" in a new design.

Using ANT for designing, speculating about making translations while maintaining social order, is not something that ANT typically does though it can be translated to that end. In this work, the insights provided by ANT about the use of ASR in the workplace have been used in the design of an ASR system for the ACT Magistrates Court.

ANT analysis of the situation I was designing for, sentencing at the Court, allowed me to see which parts could be changed and which parts *shouldn't* be changed. Then, as part of the design process for the ICAA system for the Court, parts of the ANT research frame (Callon, 1986b) were used to structure the design thinking. The most useful parts of the research frame for designing were *interresment* and *enrollment* because they specifically related to the beginning stages of establishing a new network.

8.4 Summary

Using the Locales Framework and Actor-Network Theory was useful in analysing qualitative data. Both approaches were also used in the design stage of this work. While both approaches were useful during design, neither approach replaces the creativity involved in designing. Instead they were used to guide design to begin a process of ensuring that the final product will be usable and useful.

CHAPTER 9

Conclusion

This thesis has shown how the usability of automatic speech recognition (ASR) dictation software is co-constructed through the interplay of many factors, both social and technical. Having identified these factors, this thesis has then integrated them into the preliminary design of a new ASR system for possible use at the ACT Magistrates Court.

9.1 Contributions

The contributions of this thesis are:

- The pragmatic use of existing social research methods and their antecedents as a corpus of analyses to inspire new designs;

- a demonstration of the use of Actor-Network Theory in design both as critique and as part of a design process;

- empirical field-work evidence of how large vocabulary ASR is used in the workplace;

- a design showing how ASR could be introduced to the rich, complicated, environment of the ACT Magistrates Court; and,

- a performance of the process of moving from field work to design.

Automatic speech recognition is often considered from an engineering or scientific perspective. Usability testing on ASR applications is often done in laboratory settings or as simulations of the recognition algorithm rather than long term testing in ecologically valid settings. In this thesis I have observed long-term experienced users of commercial off-the-shelf ASR applications as they used the applications in their places of work. By looking at ASR software in a naturalistic manner I have been able to move beyond considering ASR software as a stand-alone application that people work *with* to seeing it as a piece of software that people work *through*. This is an important distinction because it turns ASR from an application into an interface that mediates a user's interactions with the applications, tasks or people they need to use in their work. If the ASR application does not effectively communicate with the user and their applications then the user's work is made harder. In much of the literature on ASR usability the focus is on recognition accuracy. By observing users of ASR software in real work settings it emerged that the user interface was quite good and that the difficulty in using ASR software productively was in the ASR/application interface. The main problem in using ASR software is not one of recognition rate but of integration with the user's work process.

Going from fieldwork to the analysis summarised above required the use of analytic approaches. The approaches used in this work were the Locales Framework and Actor-Network Theory. Locales is explicitly directed at the analysis and design of computer systems where ANT originates in science and technology studies and is typically used for analysis of existing and historical socio-technical situations.

In this thesis both of these approaches were used pragmatically to ask and answer questions during the analysis of the fieldwork and during the design of the ASR system for the Court. Doing analysis of fieldwork for design is problematic. Traditional sociological analysis of fieldwork has described how a situation is or how a situation came to be. Making the leap to design inspired by fieldwork is problematic because design is about imagining the future. Both Locales and ANT were used to structure the design thinking that was inspired by the analysis of the fieldwork.

Additionally, the literature on ANT and Locales was used as a corpus of analyses that were used to direct this work's analysis and design.

The use of ANT in this thesis as a way to structure design thinking and imagine the future based on the present is novel and shows that ANT can be extended beyond analysis into design. In showing how ANT can be used as more than a tool for performing critiques of the past and present this thesis has opened up ANT to the design field. Chapters 7 and 8 show how ANT has been used in this thesis as a framework for critique and how ANT can be used in design.

Performing Locales and ANT-inspired analysis of the fieldwork described in this thesis led to the formulation of a "trajectory of use" that began with deciding to use ASR software, leaping into using it and finally arriving at a point where the sustained use requires effort to be maintained. In this concept of the use of ASR software, use is seen as a constant negotiation and balancing of competing and potentially destructive forces. By viewing ASR systems a tool that enables work to be done, it is possible to foreground the work needed to withstand many of these forces.

Finally, the insights that ASR systems are tools that enable work to be done and that use of such systems requires constant effort were illustrated in the description of a design for an ASR system at the Court. The primary reason for considering the design of a new software system for sentencing at the Court was to reduce the amount of time the magistrates spend on writing out sentences. Because the degree of fit between the work process and the capabilities of the ASR software is very important to productive use, a close examination of the work process of the Court was undertaken. The sentencing work process was examined from the point of view of the Magistrate, the Magistrate's Associate, the List Clerk and the back room workers. The fieldwork showed that the process of writing sentences on the bench was about communicating sentences to the other workers in the Court and beyond.

Analysing the fieldwork at the Court from an ASR design perspective showed where there were opportunities for design to change aspects of the Court and its work and where change could be damaging. The resulting design is not intended to be production ready, though it is based on achievable technology. Instead the design illustrates the application of the principles arrived at in the analysis of the Court and the use of ASR software. It could also be the starting point for product development.

9.2   Implications for the Design of Large Vocabulary Automatic Speech Recognition Systems

My research has implications for the design of large vocabulary ASR systems which are also known as dictation systems.

The primary goal of much ASR research is an increase in recognition accuracy. Often, an increase in the recognition accuracy of ASR software is said to make the software more usable. It is an implicit assumption that once — if — ASR software reaches 100% recognition accuracy that it will become perfectly usable. My research has shown that with a recognition rate of less than 100%, the match between the ASR software's capabilities and the work that the user is doing is very important for productive use. When users of ASR software try to do work that is beyond the capabilities of the software they are unable to work productively.

The capabilities of ASR software can be considered in two dimensions. First, and most obviously, ASR software must recognise speech. In this dimension, recognition accuracy is important. The second dimension is the match between the capabilities of the software and the user's work process. In this thesis I have considered the second dimension. Creating a match between ASR's capabilities and a particular work process is not a trivial matter. There are several ways to create such a match though any good match usually involves a combination of these techniques.

The first way to achieve a good fit between the capabilities of the ASR software is simple coincidence. As the capabilities of commercial off-the-shelf ASR software are limited to dictation tasks and some system control, this is often easy to achieve when the work process involves mainly brief correspondence or more generally, the composition of short documents. Typically, commercial off-the-shelf ASR software is not well equipped for navigating long documents so shorter documents are easier to work with in ASR.

My research has shown that establishing and maintaining a working ASR implementation requires a great deal of effort. This effort takes two forms–changing the work process and changing the ASR software. The two forms usually coexist though I will deal with them separately.

As ASR software has a finite set of capabilities it is necessary, if productive use is the goal, to work within the capabilities of the software. Often, this can simply involve changing existing office procedures to fit with the software. This could mean that ASR users have their duties modified to allow them to work within the software's capable range or that they must work without the software to achieve some tasks.

Often it will not be possible to change a work process sufficiently to utilise ASR software. An example of this is a work process that involves using software that cannot easily be controlled by ASR, for example graphics software or extensive spreadsheets. In this case it may be necessary for an ASR user (or prospective user) to change jobs. As a general rule, the more divergent a work process and the capabilities of ASR software, the more likely it is that users or prospective users will have to change jobs or adopt a different role in their organisation, a situation that occurred with a number of my interviewees.

While most ASR software has limited capabilities, it is possible to make some modifications to the software to more closely align those capabilities with the user's work process. In my observations this often took the form of the user, or another person, writing macros for the ASR software. In the Hansard office macros are used extensively to allow the Hansard editors to speak as few commands as possible and focus on their main task of transcribing what is said by the parliamentarians. At other locations I observed that users have macros to aid in the integration of their ASR software to other software they have to use in the course of their work.

The implication that this use of macros has for ASR software is that it must become easier to adapt ASR software to other software. Macros must become easier to write through the provision of more interfaces for integration and through better macro-writing software. The general public should not be expected to write code, or even pseudo-code to write simple macros to perform simple integration tasks. Visual macro recorders or automatic macro recorders need to be included with ASR software.

Where an ASR user is an occasional or short-term user, integration between software and task is not as important as where the user is a professional user of the software, working with it all day. The need for mechanisms for integration and a recognition of the importance of adapting work practices to the software is highly important for professional users and users who would need to become professional users to achieve a reasonable level of productivity. I have studied professional users of ASR software, so my findings are necessarily skewed in the direction of professional use. While large vocabulary ASR software remains errorful a focus on professional use is important because dilettante users are less likely to demand the changes necessary to improve the usefulness of ASR software that professional users demand.

By re-figuring ASR software as something users work *through* and not work *with*, it is easier to see the need for a match between the capabilities of ASR software and the tasks users are doing through it. Integrating ASR software with the software that users need to perform their work through macros and other means will improve the usability of ASR software.

In summary, the implications from my research for the design of large vocabulary ASR systems are:

- As the recognition rate approaches 100% the degree of fit between the work being done through the ASR software and the capabilities of the software becomes more important for productive use; and,

- Integration of ASR software with other software becomes essential as productive use in a real work environment becomes important.

9.3   Areas for Future Research

The contributions of this thesis to the fields of human-computer interaction and automatic speech recognition have not exhausted the work that can be done where these areas overlap. Work could also be done to extend the methodology described in this thesis to areas other than ASR. In particular, the use of Actor-Network Theory in design bith as critique and as inspiration for future designs is an area that is deserving of more research.

9.3.1 Automatic Speech Recognition

More research needs to be done on how people are integrating ASR technology into their work. It is obvious that people use ASR technology in their work and that they are successful in integrating it into their work practice, however this thesis has only opened the door to more work in this area. Future research could be done to further explain how the integration takes place and to identify in much greater detail the processes that occur to allow integration to happen.

Automatic speech recognition systems are particularly interesting because they cross a human/non-human boundary. Until automatic speech recognition systems existed, only people and some animals could understand speech. Nass and Gong (2000) have illustrated this point to some extent in showing that naive users of ASR systems do treat ASR systems as if they would understand all the nuances of speech. Further research could be done with experienced users of speech systems to determine how they approach ASR systems.

Finally, work could be done at locations or with people who are new to dictation systems who want or need to use them. The approach to the successful use of such systems described in chapter 6 certainly needs validating and refining in the workplace.

9.3.2 Other Technologies

Because I see ASR systems as a subset of a larger class of off-the-desktop systems that includes pervasive and ubiquitous computing, I see this thesis as opening areas for research with technologies of that type. Many off-the-desktop systems are based on probabilistic and error-prone techniques, as ASR is, which makes them difficult to understand and to prototype.

Biometric systems are based on similar technologies and techniques to ASR and, as such, are error-prone. As biometric systems become more widely adopted their use and misuse will become more of an issue, both in the wider community and among software professionals. The use of biometric systems in the workplace is fraught with difficulty. Methods similar to those used in this thesis could be used to explore the use of existing biometric systems as inspiration for the design of future systems. Similarly, research into biometric systems is needed that shows how such systems are integrated successfully into work so that the properties of successful implementations may be used elsewhere.

9.3.3 Actor-Network Theory in Design

Doing design inspired by fieldwork is an area that requires more research. Actor-Network Theory, and its adaptations (or translations), holds particular promise for analysis of computing systems that are "in the world". This is particularly true of the ubiquitous and pervasive computing that seeks to put computers and computing devices into many areas of everyday life. Because ubiquitous computing systems are necessarily integrated into complex heterogeneous settings where the interplay of social and technical actors can determine success or failure, ANT is ideal for analysing such situations.

Designing based on ANT is more difficult because ANT was not invented with design in mind. This thesis has shown that it is possible to create a design that is inspired and guided by ANT. Where ANT is most valuable is in projects that seek to introduce a novel design into an existing situation without radically changing the situation itself. There can be many reasons why it is desirable that an existing situation be respected but nonetheless where a new technology or other design could potentially improve some part of the situation. ANT can be used to show which actors in a situation are most vulnerable and which connections between actors are the most important so that new designs can provide real benefits without unexpected consequences. Projects that use ANT-inspired design are an area where more work should be done, both to expand the corpus of work using ANT-inspired design and also to expand the techniques used to do such design.

APPENDIX A

Application of Socially-Inspired Design for Automatic Speech Recognition

This appendix describes how to apply the methodology used in this thesis to investigate a situation where automatic speech recognition (ASR) could be useful and to apply those findings to a design for ASR in that situation. The steps for performing this methodology are described below.

1. A novel way of working at an existing location is postulated.

2. Contemporary versions of the future way of working are identified and studied for their essential properties. These may be treated as breaching experiments ((see Crabtree, 2004)).

3. The work practices of the existing situation are examined and analysed, making the practical actions clear and allowing the work to be treated as a perspicuous setting ((see Crabtree, 2004)).

4. The outcomes of step 2 and step 3 are treated as input to a design process.

5. A novel design emerges.

These steps are quite general because the actual detail of the process involved will be unique to each situation. What follows below is a more detailed examination of the five steps described above.

Having arrived at a design, which should be a description of what a new system is like to use, the construction of the new system can take place.

A.1   A Novel Way of Working

A novel way of working at an existing location is postulated.

Identifying a novel way of working is possibly the easiest step in this process. For ASR specifically it is as simple as identifying a situation in which it can be imagined that ASR will be used. It is important in this step *not* to consider how the ASR will be used or who will use it. The "way of working" is therefore quite broad in scope and must take into account the wider sphere of work than that occupied by one or two people. I would suggest starting at the organisational level and only if that is too unwieldy scoping the process to a narrower view.

In the thesis the "way of working" is the use of ASR at the ACT Magistrates Court.

A.2   Contemporary Versions of that Work

Contemporary versions of the future way of working are identified and studied for their essential properties. These may be treated as breaching experiments.

The next step is to find contemporary versions of the "way of working" identified in step one. These versions need not be exactly the same as the work identified in the first step though office work should be sought if the novel way of working is to be office based and factory work should be sought if the novel way is to be in a factory.

In the thesis the contemporary versions of the novel way of working were found with the ASR users at Hansard and in the Public Service. It was important for this thesis and indeed this process that these contemporary examples were real-world situations where people used ASR.

There will be situations that arise where no contemporary examples of work with ASR can be found to compliment the novel way of working postulated in step 1 of this process. In this case I would *suggest* that in addition to every attempt being made to locate an analogous situation that a process much more closely aligned with Crabtree's technolomethodology (Crabtree, 2004) be invoked where a situation is 'manufactured' in order to observe natural work practices with a technology.

A.3 Make the Practical Actions Clear

The work practices of the existing situation are examined and analysed, making the practical actions clear and allowing the work to be treated as a perspicuous setting

In order to break the contemporary way of working into its "essential properties" it will have to be analysed closely. Doing so is a non-trivial process and is beyond the scope of this appendix to describe. The tools for doing the analysis are introduced in section 3.2. The analysis of the contemporary ways of working identified in this thesis is described in chapter 6. Chapter 8 is valuable for anyone attempting to use this methodology because it describes my reflections on performing it with the case studies that I used.

A.4 A New Design Emerges

The outcomes of step 2 and step 3 are treated as input to a design process.

A novel design emerges.

Packaging the output of the field work for use by designers in a design process is a problem that has been tackled in many ways by many researchers. In this thesis scenarios and rich narrative descriptions are used to describe the field work. The richness of the scenarios and the inclusion of what may seem like irrelevant detail is essential to the process as the it is the richness and seemingly non-essential activities of everyday work that make the design of new ways of working so challenging. Neglecting or deliberately obfuscating minor details in the description of a work process can mean that non-procedural elements of the process that are nonetheless essential are not included in the redesigned process.

Scenarios are used in this thesis to present field work in sections 4.1 and 4.2. A rich narrative description that focuses on the use of an artefact is presented in this thesis in section 5.2. The decision to use a scenario or a narrative is a decision that is influenced by what needs to be described.

Scenarios are also used in this thesis in section 7.1 where two caricatured scenarios are presented in the style described by Bødker (2000). The methods used to generate a new design and the concepts used to think about what a new design should involve are briefly described in section 3.3. The use of these methods and concepts in this thesis is described in chapter 7. The ANT approach is elusive when described in the abstract and so the use of it in this thesis *is* the description of how to use it. The Locales Framework, being a set of tools for social analysis of situations in which technology is used is much more accessible and described well by (Fitzpatrick et al., 1998; Fitzpatrick and Kaplan, 1998; Fitzpatrick, 2002, 2003).

A.5   Construction

The methods for the construction of any system that is the result of this process need not be specified here. This methodology deals with the description of a new system in use and does not attempt to describe how such a system should be implemented.

Bibliography

Abdelnour Nocera, J. L. and Hall, P. (2004). Global software, local voices. In F. Sudweeks and C. Ess (editors), *Cultural attitudes towards communication and technology 2004*, pages 29–42. Karlstad, Sweden.

Abowd, G. D. and Mynatt, E. D. (2000). Charting past, present, and future research in ubiquitous computing. *ACM Trans. Comput.-Hum. Interact.*, 7(1):29–58. ISSN 1073-0516. doi:http://doi.acm.org/10.1145/344949. 344988.

Ahmer, I. and King, R. W. (1998). Automated captioning of television programs: development and analysis of a soundtrack corpus. In *Proceedings of ICSLP 1998*. Sydney, Australia.

Akrich, M. (1992). The de-scription of technical objects. In W. Bijker and J. Law (editors), *Shaping technology/building society*, pages 205 –224. MIT Press.

Ando, A., Kobayashi, A., and Imai, T. (1998). A thesaurus-based statistical language model for broadcast news transcription. In *Proceedings of ICSLP 1998*. Sydney, Australia.

Anthes, G. H. (2004). Speak easy. *Computerworld*, 38(27):21.

Antiles, S. E., Hornberger, C. K., Weis, M. R., Speziale, J., Bramson, R. T., and Treves, S. (2004). Implementing speech recognition at Childrens Hospital Boston. *Decisions in Imaging Economics*. Retrieved from the World Wide Web on January 25 2006 from http://www. imagingeconomics.com/library/200405-19.asp.

Bednarz, A. (2004). New speech technologies making noise. *Network World*, page 14.

Bentley, R., Hughes, J. A., Randall, D., Rodden, T., Sawyer, P., Shapiro, D., and Sommerville, I. (1992). Ethnographically-informed systems design for air traffic control. In *Proceedings of the Conference on Computer-Supported Cooperative Work*, pages 123–129. ACM Press.

Berg, S., Taylor, A. S., and Harper, R. (2003). Mobile phones for the next generation: device designs for teenagers. In *CHI '03: Proceedings of the SIGCHI conference on Human factors in computing systems*, pages 433–440. ACM Press. ISBN 1-58113-630-7. doi:http://doi.acm.org/10.1145/642611.642687.

Blomberg, J., Giacomi, J., Mosher, A., and Swenton-Wall, P. (1993). *Ethnographic field methods and their relation to design*, chapter 7. Lawrence Erlbaum Associates.

Blythin, S., Rouncefield, M., and Hughes, J. A. (1997). Never mind the ethno' stuff, what does all this mean and what do we do now: ethnography in the commercial world. *interactions*, 4(3):38–47. ISSN 1072-5520. doi:http://doi.acm.org/10.1145/255392.255400.

Bødker, S. (2000). Scenarios in user-centred design—setting the stage for reflection and action. *Interacting with Computers*, 13:61–75.

Bolt, R. A. (1980). Put-that-there: Voice and gesture at the graphics interface. In *SIGGRAPH '80: Proceedings of the 7th annual conference on Computer graphics and interactive techniques*, pages 262–270. ACM Press, New York, NY, USA. ISBN 0-89791-021-4.

Bruner, D. M. (2005). Social relations in the modern tribe: Product selection as symbolic markers. In *Proceedigns of EPIC: Ethnographic Praxis in Industry Conference*. Redmond, WA, USA.

Buckley, W. M. (2002). Why stenotypists are tying to keep Paula O'Reagan quiet. *Wall Street Journal*. June 3, 2002. Retrieved from 4 October from Factiva Database.

Buskirk, R. V. and LaLomia, M. (1995). The just noticeable difference of speech recognition accuracy. In *CHI '95: Conference companion on Human factors in computing systems*, page 95. ACM Press, New York, NY, USA. ISBN 0-89791-755-3. doi:http://doi.acm.org/10.1145/223355.223446.

Button, G. and Dourish, P. (1996). Technomethodology: paradoxes and possibilities. In *Proceedings of the SIGCHI conference on Human factors in computing systems*, pages 19–26. ISBN 0-89791-777-4.

Callon, M. (1986a). The sociology of an actor-network: The case of the electric vehicle. In M. Callon, J. Law, and A. Rip (editors), *Mapping the dynamics of science and technology: sociology of science in the real world*, chapter 2, pages 19–34. Macmillian, Houndsmills.

Callon, M. (1986b). Some elements of a sociology of translation: domestication of the scallopes and the fishermen of St Brieuc Bay. In J. Law (editor), *Power, Action and Belief*, chapter 10, pages 196–233. Routledge & Kegan Paul, London, Boston and Henley.

Campbell, B., Cederman-Haysom, T., Donovan, J., and Brereton, M. (2003). Springboards into design: Exploring multiple representations of interaction in a dental surgery. In Viller and Wyeth (editors), *Proceedings of OzCHI2003*. CHISIG, Brisbane, Australia.

Carroll, J. (editor) (1995). *Scenario-based design. Envisioning work and technology in System Development*. Wiley, New York.

Clark, D. (2001). Speech recognition: The wireless interface revolution. *Computer*, 34(3):16–18. ISSN 0018-9162.

Clarke, A. C. (1974). *Profiles of the future : an inquiry into the limits of the possible*. Gollancz, London.

Cohen, P. R. and Oviatt, S. L. (1995). The role of voice input for human-machine communication. In *Proceedings of the National Academy of Sciences of the United States of America*, volume 92, pages 9921–9927.

ComCare Website (2005). About ComCare. http://www.comcare. gov.au/overview.html. Retreived on 20 January, 2005.

Cook, G., Robinson, T., and Christie, J. (1998). Real-time recognition of broadcast news. In *Proceedings of ICSLP 1998*. Sydney, Australia.

Court of the Future website, 2004 (2004). National court of the future. http://ncf.canberra.edu.au/indexf.html. Retrieved from the world wide web on 1 June 2004.

Crabtree, A. (2004). Taking technomethodology seriously: hybrid change in the ethnomethodology-design relationship. *Eur. J. Inf. Syst.*, 13(3):195–209. ISSN 0960-085X. doi:http://dx.doi.org/10.1057/palgrave.ejis. 3000500.

Crabtree, A., Nichols, D. M., O'Brien, J., Rouncefield, M., and Twidale, M. B. (2000). Ethnomethodologically informed ethnography and information system design. *Journal of the American Society for Information Science*, 51(7):666–682.

Danis, C., Comerford, L., Janke, E., Davies, K., Vries, J. D., and Bertrand, A. (1994). Storywriter: a speech oriented editor. In *CHI '94: Conference companion on Human factors in computing systems*, pages 277–278. ACM Press, New York, NY, USA. ISBN 0-89791-651-4. doi:http://doi.acm. org/10.1145/259963.260490.

Danis, C. and Karat, J. (1995). Technology-driven design of speech recognition systems. In *DIS '95: Proceedings of the conference on Designing Interactive Systems*, pages 17–24. ACM Press, New York, NY, USA. ISBN 0-89791-673-5. doi:http://doi.acm.org/10.1145/225434.225437.

Dourish, P. (2001). *Where the action is: the foundations of embodied interaction*. MIT Press, Cambridge, Mass.

Dourish, P. and Button, G. (1998). On "technomethodology": Foundational relationships between ethnomethodology and system design. *Human Computer Interaction*, 13(4):395–432.

Dragon NaturallySpeaking Preferred 6 (2002). Dragon NaturallySpeaking Preferred 6. Software box.

Feng, J. (2002). Improving speech-based navigation during dictation. In *CHI '02: CHI '02 extended abstracts on human factors in computing systems*, pages 844–845. ACM Press, New York, NY, USA. ISBN 1-58113-454-1. doi:http://doi.acm.org/10.1145/506443.506627.

Feng, J. and Sears, A. (2004). Using confidence scores to improve hands-free speech based navigation in continuous dictation systems. *ACM Trans. Comput.-Hum. Interact.*, 11(4):329–356. ISSN 1073-0516. doi:http://doi.acm.org/10.1145/1035575.1035576.

Fitzpatrick, G. (2002). The locales framework: making social thinking accessible for software practitioners. In R. Klischewski, C. Floyd, and Y. Dittrich (editors), *Social thinking, software practice*. MIT Press.

Fitzpatrick, G. (2003). *The Locales Framework: Understanding and designing for wicked problems*. Kluwer Academic Publishers.

Fitzpatrick, G. and Kaplan, S. (1998). Applying the locales framework to understanding and designing. In P. Calder and B. Thomas (editors), *Proceedings of Ozchi 1998*. IEEE.

Fitzpatrick, G., Kaplan, S., and Mansfield, T. (1998). Applying the locales framework to understanding and designing. In *OZCHI '98: Proceedings of the Australasian Conference on Computer Human Interaction*, page 122. IEEE Computer Society. ISBN 0-8186-9206-5.

Fox, P. (2003). Let's talk (about speech recognition). *Computerworld*, 37(47):20.

Gauvain, J.-L., Lamel, L., and Adda, G. (1998). Partitioning and transcription of broadcast news data. In *Proceedings of ICSLP 1998*. Sydney, Australia.

Gibson, W. (1995). *Neuromancer*. Voyager, London. First published 1984.

Gibson, W. (1996). *Idoru*. Penguin, England.

Glasser, B. and Strauss, A. (1967). *The Discovery of Grounded Theory*. Aldine Publishing Co., Chicago.

Goette, T. (2000). Keys to the adoption and use of voice recognition technology in organizations. *Information Technology & People*, 13(1).

Gomm, K. (2004). Grocery wholesaler reduces stock errors to almost zero with voice recognition system. *Computer Weekly*, page p12.

Goronzy, S. and Beringer, N. (2005). Integrated development and on-the-fly simulation of multimodal dialogs. In *Proceedings of Interspeech 2005*. Lisboa.

Gould, J. D. (1978). How experts dictate. *Journal of Experimental Psychology: Human Perception and Performance.*, 4(4):648–661.

Gould, J. D., Conti, J., and Hovanyecz, T. (1983). Composing letters with a simulated listening typewriter. *Commun. ACM*, 26(4):295–308. ISSN 0001-0782. doi:http://doi.acm.org/10.1145/2163.358100.

Graham, C., Cheverst, K., and Rouncefield, M. (2005). Technology for the humdrum: trajectories, interactional needs and a care setting. In *OZCHI '05: Proceedings of the 19th conference of the computer-human interaction special interest group (CHISIG) of Australia on Computer-human interaction*, pages 1–10. Computer-Human Interaction Special Interest Group (CHISIG) of Australia, Narrabundah, Australia. ISBN 1-59593-222-4.

Grimm, J. and Grimm, W. (1975). *Snow White and the Seven Dwarfs*. Larousse. Ilustrated by Otto S. Svend. Translated by Anne Rogers.

Halstead-Nussloch, R. (1989). The design of phone-based interfaces for consumers. In *CHI '89: Proceedings of the SIGCHI conference on Human factors in computing systems*, pages 347–352. ACM Press, New York, NY, USA. ISBN 0-89791-301-9. doi:http://doi.acm.org/10.1145/67449.67516.

Halverson, C., Horn, D., Karat, C., and Karat, J. (1999). The beauty of errors: patterns of error correction in desktop speech system. In *Proceedings of INTERACT '99*, pages 133–140. IOS Press, Edinburgh.

Hanseth, O. and Monteiro, E. (1998). Understanding information infrastructure. http://heim.ifi.uio.no/~oleha/Publications/bok.html. Retrieved on 28 January, 2006.

Heath, C. and Luff, P. (1992). Collaboration and control: Crisis management and multimedia technology in london underground line control rooms. *Computer Supported Cooperative Work*, 1(1-2):69–94.

Heisterkamp, P. (2001). Linguatronic product-level speech system for mercedes-benz cars. In *HLT '01: Proceedings of the first international conference on Human language technology research*, pages 1–2. Association for Computational Linguistics, Morristown, NJ, USA.

Honda Motor Company Ltd. (2004). A car that listens to you; IBM and Honda say they've developed a hands-free, natural-sounding speech-recognition system for advanced navigation in cars. *InformationWeek*.

Huang, X., Acero, A., Alleva, F., Hwang, M., Jiang, L., and Mahajan, M. (1999). From Sphinx-II to whisper—making speech recognition usable. In C.-H. Lee, F. K. Soong, and K. K. Paliwal (editors), *Automatic Speech and Speaker Recognition: Advanced Topics*, chapter 20, pages 481–508. Kluwer Academic Publishers, Norwell, MA, USA.

Huber, B. (2000). Raging against the machine. *Journal of Court reporting*. Retrieved from the World Wide Web on September 28 2002 http://www.ncraonline.org/jcr/0011/0011_02.htm.

Hughes, J., King, V., Rodden, T., and Andersen, H. (1995). The role of ethnography in interactive systems design. *interactions*, 2(2):56–65. ISSN 1072-5520. doi:http://doi.acm.org/10.1145/205350.205358.

Hughes, J., Randall, D., and Shapiro, D. (1992a). Designing with ethnography: Making work visible. *Interacting with Computers*, 5(2).

Hughes, J. A., Randall, D., and Shapiro, D. (1992b). Faltering from ethnography to design. In *CSCW '92: Proceedings of the 1992 ACM conference on Computer-supported cooperative work*, pages 115–122. ACM Press. ISBN 0-89791-542-9. doi:http://doi.acm.org/10.1145/143457.143469.

James, F., Lai, J., Suhm, B., Balentine, B., Makhoul, J., Nass, C., and Shneiderman, B. (2002). Getting real about speech: overdue or overhyped? In
CHI '02: CHI '02 extended abstracts on Human factors in computing systems,
pages 708–709. ACM Press, New York, NY, USA. ISBN 1-58113-454-1.
doi:http://doi.acm.org/10.1145/506443.506557.

JCR Online (2000). Speech recogntion and court reporting. *Journal of
Court Reporting*. Retrieved from the World Wide Web on September
28 2002 from http://www.verbatimreporters.com/jcr/0011/
0011_03.htm.

JCR Online (2001). Interview: An update on voice writer realtime reporting. *Journal of Court Reporting*. Retrieved from the World Wide Web
on September 28 2002 from http://www.ncraonline.org/jcr/
0101/0101_interview.htm.

Jordan, K. and Lynch, M. (1992). The sociology of a genetic engineering
technique: ritual and rationality in the performance of the "plasmid
prep". In A. E. Clarke and J. H. Fujimura (editors), *The right tools for
the right job: a work in twentieth-century life sciences*, chapter 3. Princeton
University Press.

Kaku, M. (1998). *Visions: How science will revolutionise the 21st century and
beyond*. Oxford University Press.

Karat, C.-M., Halverson, C., Horn, D., and Karat, J. (1999). Patterns of
entry and correction in large vocabulary continuous speech recognition
systems. In *CHI '99: Proceedings of the SIGCHI conference on Human factors
in computing systems*, pages 568–575. ACM Press, New York, NY, USA.
ISBN 0-201-48559-1. doi:http://doi.acm.org/10.1145/302979.303160.

Karat, J., Horn, D. B., Halverson, C. A., and Karat, C. M. (2000). Overcoming unusability: developing efficient strategies in speech recognition systems. In *CHI '00: CHI '00 extended abstracts on Human factors
in computing systems*, pages 141–142. ACM Press, New York, NY, USA.
ISBN 1-58113-248-4. doi:http://doi.acm.org/10.1145/633292.633372.

Kraal, B., Collings, P., Dugdale, A., and Wagner, M. (2004). An ethnography of speech recognition. In *OZCHI 2004*. CHISIG, University of Wollongong, NSW.

Kraal, B., Wagner, M., and Collings, P. (2002). Improving the design of dictation software. In *Australian Speech Science and Technology Conference*. ASSTA, University of Melbourne, Victoria, Australia.

Lai, J. and Vergo, J. (1997). Medspeak: report creation with continuous speech recognition. In *CHI '97: Proceedings of the SIGCHI conference on Human factors in computing systems*, pages 431–438. ACM Press, New York, NY, USA. ISBN 0-89791-802-9. doi:http://doi.acm.org/10.1145/258549.258829.

Lane, B. A. (1988). Elementary, Dear Data. http://www.st-minutiae.com/academy/literature329/129.txt. Retrieved on 1 December, 2004.

Latour, B. (1987). *Science in action : how to follow scientists and engineers through society*. Harvard University Press, Cambridge, Mass.

Latour, B. (1992). Where are the missing masses? In W. Bijker and J. Law (editors), *Shaping technology/ building society*, pages 205 –224. MIT Press.

Latour, B. (1996). *Aramis, or, The love of technology*. Harvard University Press, Cambridge, Mass.

Law, J. (1986). On methods of long-distance control: vessels, navigation and the Portuguese route to India. In J. Law (editor), *Power, Action and Belief*, chapter 10, pages 196–233. Routledge & Kegan Paul, London, Boston and Henley.

Law, J. (1987). Technology and heterogeneous engineering: The case of the Portuguese expansion. In W. Bijker, T. Hughes, and T. Pinch (editors), *The Social Construction of Technological Systems*. MIT Press, Cambridge, Ma.

Law, J. (2003). Traduction/trahison: Notes on ant. `http://www.lancs.ac.uk/fss/sociology/papers/law-traduction-trahison.pdf%`. Retrieved from the World Wide Web on Monday July 11, 2005.

Law, J. and Hassard, J. (editors) (1999). *Actor network theory and after*. Blackwell Publishers, Oxford.

Ligget, W. and Fisher, W. (1998). Insights from the broadcast news benchmark tests. In *DARPA Broadcast News Transcription and Understanding Workshop*, pages 16–22.

Lippmann, R. P. (1997). Speech recognition by machines and humans. *Speech Communication*, 22:1–15.

Lockwood, P. and Boudy, J. (1992). Experiments with a nonlinear spectral subtractor (nss), hidden markov models and the projection, for robust speech recognition in cars. *Speech Commun.*, 11(2-3):215–228. ISSN 0167-6393. doi:http://dx.doi.org/10.1016/0167-6393(92)90016-Z.

Macinnes, B. (2004). Convergence expected to drive speech recognition. *MicroScope*, page p16.

March, W. and Fleuriot, C. (2005). The worst technology for girls? In *Proceedigns of EPIC: Ethnographic Praxis in Industry Conference*. Redmond, WA, USA.

Meadors, J., Kanabay, H. D., Starkman, I., and Carroll, B. (2001). Voice writers as members: Two opinions. *Journal of Court Reporting*, February 2001. Retrieved from the World Wide Web on September 28 2002 from `http://www.ncraonline.org/meetings/ac2001/0102_viewpoints.htm`.

Moore, R. K. (2004). Modelling data entry rates for ASR and alternative input methods. In *Proceedings of INTERSPEECH 2004*. Korea.

Murray, C. J. (2004). Talking in-car navigation system ups tech ante. *Electronic Engineering Times*, page p14.

Nanavati, A. A. and Rajput, N. (2005). Characterising dialogue call-flows for pervasive environments. In *Proceedings of Interspeech 2005*. Lisboa.

Nardi, B. A. and Miller, J. R. (1990). An ethnographic study of distributed problem solving in spreadsheet development. In *CSCW '90: Proceedings of the 1990 ACM conference on Computer-supported cooperative work*, pages 197–208. ACM Press, New York, NY, USA. ISBN 0-89791-402-3. doi: http://doi.acm.org/10.1145/99332.99355.

Nass, C. and Gong, L. (2000). Speech interfaces from an evolutionary perspective. *Commun. ACM*, 43(9):36–43. ISSN 0001-0782. doi: http://doi.acm.org/10.1145/348941.348976.

Nielsen, J. (1993). *Usability engineering*. Academic Press, Boston.

Norman, D. A. (1988). *The psychology of everyday things*. Basic Books, New York.

Oria, D. and Koskinen, E. (2002). E-mail goes mobile: the design and implementation of a spoken language interface to e-mail. In *Proceedings of ICSLP 2002*. Denver, Colorado.

Orlikowski, W. J. and Gash, D. C. (1994). Technological frames: making sense of information technology in organizations. *ACM Trans. Inf. Syst.*, 12(2):174–207. ISSN 1046-8188. doi:http://doi.acm.org/10.1145/196734.196745.

Orwill, G. (2003). *Nineteen eighty-four : a novel*. Plume, New York.

Oviatt, S. (1999a). Mutual disambiguation of recognition errors in a multimodel architecture. In *CHI '99: Proceedings of the SIGCHI conference on Human factors in computing systems*, pages 576–583. ACM Press, New York, NY, USA. ISBN 0-201-48559-1. doi:http://doi.acm.org/10.1145/302979.303163.

Oviatt, S. (1999b). Ten myths of multimodal interaction. *Commun. ACM*, 42(11):74–81. ISSN 0001-0782. doi:http://doi.acm.org/10.1145/319382.319398.

Oviatt, S., Coulston, R., Tomko, S., Xiao, B., Lunsford, R., Wesson, M., and Carmichael, L. (2003). Toward a theory of organized multimodal integration patterns during human-computer interaction. In *ICMI '03: Proceedings of the 5th International Conference on Multimodal Interfaces*, pages 44–51. ACM Press, New York, NY, USA. ISBN 1-58113-621-8. doi: http://doi.acm.org/10.1145/958432.958443.

Pennington, M. (2002). More than just a court reporter. *Journal of Court Reporting*. Retrieved September 28 2002 from the World Wide Web http://www.ncraonline.org/jcr/0210/0210_profession.pdf.

Polansky, L. G. (1997). Speech recognition - the court technology for the 21st century. In *Fifth National Court Technology Conference (CTC5*. National Center for State Courts. Retrieved 28 September 2002 from the World Wide Web http://www.ncsconline.org/D_Tech/CTC/CTC5/302.HTM.

Prasad, R., Nguyen, L., Schwartz, R., and Mankoul, J. (2002). Automatic transcription of courtroom speech. In *Proceedings of the 7th International Conference on Spoken Language Processing*, volume 3, pages 1745–1748. Denver, Colorado, USA.

Randall, D., Harper, R., and Rouncefield, M. (2005). Fieldwork and ethnography: A perspective from CSCW. In *Proceedigns of EPIC: Ethnographic Praxis in Industry Conference*. Redmond, WA, USA.

Read, J. C., MacFarlane, S., and Casey, C. (2002). Oops! Silly me! Errors in a handwriting recognition-based text entry interface for children. In *NordiCHI '02: Proceedings of the second Nordic conference on Human-computer interaction*, pages 35–40. ACM Press, New York, NY, USA. ISBN 1-58113-616-1. doi:http://doi.acm.org/10.1145/572020.572026.

Reed Business Information Ltd. (2004). Speech recognition technology reaches 95% accuracy. *Electronics Weekly*, page p1.

Robert-Ribes, J. (1998). On the use of automatic speech recognition for TV captioning. In *Proceedings of ICSLP 1998*. Sydney, Australia.

Robertson, T. (2002). The public availability of actions and artefacts. *Computer Supported Cooperative Work: The Journal of Collaborative Computing*, 11(3-4):299–316.

Robertson, T. J. (1997). *Designing Over Distance*. Ph.D. thesis, School of Computing Sciences, University of Technology, Sydney.

Rogers, Y. and Bellotti, V. (1997). Grounding blue-sky research: how can ethnography help? *interactions*, 4(3):58–63. ISSN 1072-5520. doi:http://doi.acm.org/10.1145/255392.255404.

Rollins, A., Constantine, B., and Baker, S. (1983). Speech recognition at two field sites. In *CHI'83 Proceedings*, pages 267–273.

Rosenfeld, R., Olsen, D., and Rudnicky, A. (2000a). Universal human-machine speech interface. White Paper CMU-CS-00-114, Carnegie Mellon University.

Rosenfeld, R., Olsen, D., and Rudnicky, A. (2001). Universal speech interfaces. *interactions*, 8(6):34–44. ISSN 1072-5520. doi:http://doi.acm.org/10.1145/384076.384085.

Rosenfeld, R., Zhu, X., Toth, A., Shriver, S., Lenzo, K., and Black, A. W. (2000b). Towards a universal speech interface. In *Proceedings of the International Conference on Spoken Language Processing*. Beijing, China.

Rouncefield, M., Highes, J. A., Rodden, T., and Viller, S. (1994). Working with "constant interruption": CSCW and the small office. In *Proceedings of CSCW 94*, pages 275–286. Chapel Hill, NC, USA.

Satchell, C. (2003). The swarm: Facilitating fluidity and control in young people's use of mobile phones. In Viller and Wyeth (editors), *Proceedings of OzCHI2003*. CHISIG, Brisbane, Australia.

Scansoft (2005). Scansoft. http://www.scansoft.com.

Shneiderman, B. (1992). *Designing the User Interface*. Addison-Wesley, second edition.

Shneiderman, B. (2000). The limits of speech recognition. *Commun. ACM*, 43(9):63–65. ISSN 0001-0782. doi:http://doi.acm.org/10.1145/348941. 348990.

Shriver, S. and Rosenfeld, R. (2002). Keywords for a universal speech interface. In *CHI '02: CHI '02 extended abstracts on Human factors in computing systems*, pages 726–727. ACM Press, New York, NY, USA. ISBN 1-58113-454-1. doi:http://doi.acm.org/10.1145/506443.506567.

Shriver, S., Rosenfeld, R., Zhu, X., Toth, A., Rudnicky, A., and Flueckiger, M. (2001). Universalizing speech: Notes from the USI project. In *Proceedings of Eurospeech '01*. Aalborg, Denmark.

Simonsen, J. and Kensing, F. (1997). Using ethnography in contextual design. *Commun. ACM*, 40(7):82–88. ISSN 0001-0782. doi:http://doi.acm. org/10.1145/256175.256190.

Snape, L., Casey, C., MacFarlane, S. J., and Robertson, L. (1997). Using speech in multimedia applications. In *TCS Conference*. Bangor, Wales.

Snyder, C. (2003). *Paper prototyping : the fast and easy way to design and refine user interfaces*. Morgan Kaufmann, San Francisco, Calif.

Soto, C. A. (2004). NaturallySpeaking dictates performance. *Government Computer News*, 23(1):31.

Squires, S. and Byrne, B. (editors) (2002). *Creating breakthrough ideas: the collaboration of anthropologists and designers in the product development industry*. Bergin & Garvey, Westport, Conn.

Stephenson, N. (1992). *Snow Crash*. Penguin, London , England.

Strauss, A. (1993). *Continual Permutations of Action*. Aldine De Gruyter, New York.

Suchman, L. (1987). *Plans and situated action*. Cambridge University Press.

Suchman, L. (2003). Anthropology as 'brand': Reflections on corporate anthropology. Retrieved from the World Wide Web on November

30, 2005, from `http://comp.lancs.ac.uk/sociology/papers/Suchman-Anthropology-as-Brand.%pdf`.

Suhm, B., Myers, B., and Waibel, A. (2001). Multimodal error correction for speech user interfaces. *ACM Trans. Comput.-Hum. Interact.*, 8(1):60–98. ISSN 1073-0516. doi:http://doi.acm.org/10.1145/371127.371166.

Tatnall, A. and Gilding, A. (1999). Actor-network theory and information systems research. In *Proceedings of the 10th Australasian Conference on Information Systems*. Wellington, New Zealand.

Taylor, A. S. and Harper, R. (2002). Age-old practices in the 'new world': a study of gift-giving between teenage mobile phone users. In *CHI '02: Proceedings of the SIGCHI conference on Human factors in computing systems*, pages 439–446. ACM Press. ISBN 1-58113-453-3. doi: http://doi.acm.org/10.1145/503376.503455.

Taylor, A. S. and Harper, R. (2003). The gift of the gab?: A design oriented sociology of young people's use of mobiles. *Computer Supported Cooperative Work*, 12(3):267–296. ISSN 0925-9724. doi:http://dx.doi.org/10.1023/A:1025091532662.

Van Buskirk, R. and LaLomia, M. (1995). The just noticeable difference of speech recognition accuracy. In *Conference companion on Human factors in computing systems*, page 95. ISBN 0-89791-755-3.

Vanhoucke, V., Neely, W. L., Mortati, M., Sloan, M. J., and Nass, C. (2001). Effects of prompt style when navigating through structured data. In M. Hirose (editor), *Proceedings of INTERACT '01*.

Wactlar, H. D., Hauptmann, A. G., and Witbrock, M. J. (1998). Infomedia news-on-demand: using speech recognition to create a digital video library. Technical report, Carnegie Mellon University. Technical Report CMU-CS-98-110.

Ward, N. G., Rivera, A. G., Ward, K., and Novick, D. G. (2005). Root causes of lost time and user stress in a simple dialog system. In *Proceedings of Interspeech 2005*. Lisboa.

Weinschenk, S. and Barker, D. T. (2000). *Designing effective speech interfaces.* J. Wiley, New York.

Whittaker, S. and Amento, B. (2004). Semantic speech editing. In *CHI '04: Proceedings of the SIGCHI conference on Human factors in computing systems*, pages 527–534. ACM Press, New York, NY, USA. ISBN 1-58113-702-8. doi:http://doi.acm.org/10.1145/985692.985759.

Whittaker, S., Hirschberg, J., Amento, B., Stark, L., Bacchiani, M., Isenhour, P., Stead, L., Zamchick, G., and Rosenberg, A. (2002). Scanmail: a voicemail interface that makes speech browsable, readable and searchable. In *CHI '02: Proceedings of the SIGCHI conference on Human factors in computing systems*, pages 275–282. ACM Press. ISBN 1-58113-453-3. doi:http://doi.acm.org/10.1145/503376.503426.

Whittaker, S., Hirschberg, J., Choi, J., Hindle, D., Pereira, F., and Singhal, A. (1999). Scan: designing and evaluating user interfaces to support retrieval from speech archives. In *SIGIR '99: Proceedings of the 22nd annual international ACM SIGIR conference on Research and development in information retrieval*, pages 26–33. ACM Press. ISBN 1-58113-096-1. doi: http://doi.acm.org/10.1145/312624.312639.

Wilkie, J., Jack, M. A., and Littlewood, P. (2002). Design of system-initiated digressive proposals for automated banking dialogues. In *Proceedings of ICSLP 2002*. Denver, Colorado.

Yankelovich, N. (1996). How do users know what to say? *interactions*, 3(6):32–43. ISSN 1072-5520. doi:http://doi.acm.org/10.1145/242485.242500.

Ziff Davis Media Inc. (2004). Voice software makes strides. *PC Magazine*. Web edition.

www.ingramcontent.com/pod-product-compliance
Lightning Source LLC
Chambersburg PA
CBHW071425050326
40689CB00010B/1981